A-LEVEL YEAR 2

STUDENT GUIDE

cirencester
college
a beacon college

WJEC/Eduqas

Geography

Global systems: water and carbon cycles

Fieldwork and investigative skills

Simon Oak

HODDER
EDUCATION
AN HACHETTE UK COMPANY

Dedicated with gratitude to Roger Pickup

Hodder Education, an Hachette UK company, Blenheim Court, George Street, Banbury, Oxfordshire OX16 5BH

Orders

Bookpoint Ltd, 130 Park Drive, Milton Park, Abingdon, Oxfordshire OX14 4SE

tel: 01235 827720

fax: 01235 400401

e-mail: education@bookpoint.co.uk

Lines are open 9.00 a.m.–5.00 p.m., Monday to Saturday, with a 24-hour message answering service. You can also order through the Hodder Education website: www.hoddereducation.co.uk

ISBN 978-1-4718-6415-5

First printed 2017

Impression number 5 4

Year 2020

This guide has been written specifically to support students preparing for the WJEC/Eduqas A-level Geography examinations. The content has been neither approved nor endorsed by WJEC/Eduqas and remains the sole responsibility of the author.

Cover photo: dabldy/Fotolia; other photos p. 35 (cartoon) Giles/Express Newspapers/N&S Syndication; p. 37 © NASA Earth Observatory/MODIS/Jesse Allen; p. 53 David Chapman/ Alamy Stock Photo; p. 57 Martin Evans; pp. 74, 80 Simon Oakes

Typeset by Integra Software Services Pvt Ltd, Pondicherry, India

Printed in India

Hachette UK's policy is to use papers that are natural, renewable and recyclable products and made from wood grown in sustainable forests. The logging and manufacturing processes are expected to conform to the environmental regulations of the country of origin.

Contents

Getting the most from this book 4
About this book 5

Content Guidance

Water and carbon cycles

The concepts of system and mass balance 10
Catchment hydrology: the drainage basin as a system 15
Temporal variations in river discharge 22
Precipitation and excess runoff in the water cycle 27
Deficit in the water cycle 33
The global carbon cycle 39
Carbon stores in different biomes 45
Changing carbon stores in peatlands over time 52
Links between the water and carbon cycles 58
Feedback in and between the carbon and water cycles 63

Fieldwork and investigative skills

Planning an independent investigation 71
Carrying out an independent investigation 77
Completing an independent investigation 80

Questions & Answers

About this section 85
Sample questions 86
Knowledge check answers 98
Self-study task answers 99
Index 102

■ Getting the most from this book

Exam tips

Advice on key points in the text to help you learn and recall content, avoid pitfalls, and polish your exam technique in order to boost your grade.

Knowledge check

Rapid-fire questions throughout the Content Guidance section to check your understanding.

Knowledge check answers

1 Turn to the back of the book for the Knowledge check answers.

Summaries

■ Each core topic is rounded off by a bullet-list summary for quick-check reference of what you need to know.

Sample student answers

Practise the questions, then look at the student answers that follow.

Exam-style questions

Commentary on the questions

Tips on what you need to do to gain full marks, indicated by the icon **e**

Commentary on sample student answers

Read the comments (preceded by the icon **e**) showing how many marks each answer would be awarded in the exam and exactly where marks are gained or lost.

Questions & Answers

Student answer

At the start of the storm we can assume that the vegetation surfaces are dry and have plenty of capacity to store new inputs of rainwater, especially if the vegetation consists of broad-leaved trees. This is why there is no overland flow and virtually no throughflow to begin with. However, soon after the rain begins, the capacity of leaf surfaces to store further rainfall is exhausted. Also, water which has collected in ground surface depressions begins to soak into the soil and generate throughflow. This is why throughflow becomes the dominant process after 30 minutes. Later on in the storm event, saturation of the soil results in a rapid increase in overland flow because most further rainfall cannot be stored anywhere.

e 5 marks awarded This is a thorough account that acknowledges the main changes explicitly in addition to the changing relative importance of different processes over time. Although direct precipitation is not mentioned, all of the other main processes are dealt with in a clear and well explained way which also pays close attention to details (the comment about leaf type is good). Overall, this would achieve Band 3.

Question 2

This question follows a format used by both WJEC and Eduqas A-level short structured questions. You have around 12–13 minutes to study Table 1 and answer both parts of the question (which relates to the carbon cycle).

Table 1 Variations in carbon storage (amount and location) for selected biomes

Natural biomes	Total carbon stored globally (GtC)	Where most carbon is stored in the biome
Temperate grasslands	184	Below ground
Tropical rainforests	548	Above ground
Deserts	178	Below ground
Tundra	155	Below ground

(a) Suggest reasons for the variations in carbon storage shown in Table 1. (5 marks)

e Note the command word: this answer asks for suggested reasons (explanations) to be provided using applied knowledge (AO2). A skills-based (AO3) description of the data is *not* required. Suggested reasons might include the following explanatory points:

■ Tropical rainforest is by far the largest store (around three times higher than the others) on account of its year-long growing season and constant supply of precipitation.
■ Tundra has the lowest storage (155 GtC) because of the lack of light in winter, low temperatures and relatively low precipitation.
■ Arid deserts and grassland experience low rainfall as a limiting factor.
■ Only tropical rainforests have a climate that can support trees, hence the large above-ground biomass.

■ About this book

This guide has been designed to help you succeed in the Eduqas and WJEC Geography A-level courses. The topics and components covered in this guide are:

■ Section A of WJEC A-level Unit 3: Global systems (water and carbon cycles)
■ Section A of Eduqas A-level Component 2: Global systems (water and carbon cycles)
■ WJEC A-level Unit 5 and Eduqas A-level Component 4: The independent investigation (non-examined assessment)

The format of the different examination papers is summarised in the table below.

Specification and paper number	Total marks for Section A	Suggested time spent on Section A	Structured questions	Extended response/essay
WJEC A2 Unit 3	35/96	45 min in paper lasting for 2h	Section A has **two** compulsory, structured questions with data response marked out of 17	Section A has an essay marked out of 18 (choose from **two**)
Eduqas A-level Component 2	40/110	45 min in paper lasting for 2h	Section A has **two** compulsory, structured questions with data response marked out of 20	Section A has an essay marked out of 20 (choose from **two**)

The guide has the following main objectives:

■ It provides you with key concepts, definitions, theories and examples that may be used to answer questions in the examination. The examples have been chosen to provide you with an up-to-date view of important global issues.
■ It provides guidance on how to tackle the synoptic element of the exam, which draws on ideas from your entire geography course.
■ It suggests self-study tasks that will enhance your knowledge and understanding before you enter the examinations.
■ Finally, it gives you the opportunity to test yourself through knowledge check questions, which are designed to help you check your depth of knowledge. You will also benefit from noting the exam tips, which provide further help in determining how to learn key aspects of the course.

The examinations

Eduqas

At A-level, the total examination time of Component 2: Global systems and global governance is 2 hours. You have approximately 40–45 minutes to answer the questions in **Section A: Global systems (water and carbon cycles)**, which comprises **two** compulsory structured, data response questions and **one** extended essay question (choose from two).

The two structured questions start with a part (a) resource-based question: you may be asked to describe a pattern or trend. The resource can also lead to a question requiring basic calculation or some interpretation of the data, which might be in cartographic, graphical, statistical or photographic form (including air and satellite images). The following part (b) question develops from the topic in (a), it requires brief explanation of a relatively small section of the specification and is worth 5 marks.

The essay question that follows is worth 20 marks. You are expected to offer a structured response consisting of continuous prose. A command word or phrase such as 'discuss' or 'to what extent' requires you to adopt an evaluative approach and arrive at a conclusion. There is a choice from two titles.

WJEC

The total examination time of WJEC A-level Unit 3: Global systems and global governance is 2 hours. You have approximately 40–45 minutes to answer the questions in **Section A: Global systems (water and carbon cycles)**, which comprises **two** compulsory structured, data response questions and **one** extended essay question (choose from two).

The two structured questions start with a part (a) resource-based question: you may be asked to describe a pattern or trend, or to suggest reasons why the data appear the way they do. The resource can also lead to a question requiring basic calculation or some interpretation of the data, which might be in cartographic, graphical, statistical or photographic form (including air and satellite images). The following part (b) question develops from the topic in (a), it requires brief explanation of a relatively small section of the specification and is worth 4 or 5 marks.

The essay question that follows is worth 18 marks. You are expected to offer a structured response consisting of continuous prose. A command word or phrase such as 'discuss' or 'to what extent' requires you to adopt an evaluative approach and arrive at a conclusion. There is a choice from two titles.

How answers are marked

The examiner will use **Assessment Objectives** (**AOs**) to mark your work. The AOs for A2 (WJEC) and A-level (Eduqas) are as follows.

AO1: Demonstrate knowledge and understanding of places, environments, concepts, processes, interactions and change at a variety of scales.

AO2: Apply knowledge and understanding in different contexts to interpret, analyse and evaluate geographical information and issues.
- AO2.1a: Apply knowledge and understanding in different contexts to analyse geographical information and issues.
- AO2.1b: Apply knowledge and understanding in different contexts to interpret geographical information and issues.
- AO2.1c: Apply knowledge and understanding in different contexts to appraise/judge geographical information.

AO3: Use a variety of relevant quantitative, qualitative and fieldwork skills to:
- AO3.1: Investigate geographical questions and issues.
- AO3.2: Interpret, analyse and evaluate data and evidence.
- AO3.3: Construct arguments and draw conclusions.

Mark bands

For the AOs being tested in each question the examiner will make use of marks bands for each AO to guide his/her decision. Here are the qualities that examiners will be looking for in each final band.

Band 3: Answers to the shorter questions (4 or 5 marks) will be clear and factually accurate, displaying good knowledge and understanding supported by developed examples, sketches and diagrams. Descriptions will be clear. Statistical work will be complete and understood. Answers to the **essays** will be well written and argued so that the command word (e.g. 'discuss' or 'assess') has been followed. Knowledge will be very detailed, accurate and well supported by examples, and issues will be fully understood.

Band 2: Answers to the shorter questions are often unbalanced and partial responses, which may be unstructured and make points in a random order. Knowledge is present but not always factually accurate or completely understood. Answers to the **essays** will demonstrate some understanding but not of all of the points and the examples will be mostly accurate and rather sketchy. Diagrams and statistical work may be less complete. The command word is reinterpreted to mean description rather than discussion or evaluation. The coverage might be limited to short case studies with generalised detail.

Band 1: Answers to the shorter questions have very limited and possibly fragmented factual knowledge. There might be no valid examples or just a single example named but not developed. Statistical responses will be neglected. Answers to the **essays** may be a set of unrelated, undeveloped ideas, possibly only in note form, and rather hit and miss in their relevance to the question. The command word might be ignored.

The independent investigation

You are expected to complete an independent investigation as part of your A-level course. Also called the **non-examined assessment** (because your own teacher reads and marks your document when it is completed), the independent investigation is integral to A-level geography and contributes 20% to the overall final assessment. Your independent investigation will:

- be based on a question or issue you have identified and developed *yourself* (it can address aims, questions and/or hypotheses relating to any of the compulsory or optional content from across the entire geography specification)
- draw on your own research, including field data you have sourced by yourself, or data created as part of a group fieldtrip (providing certain rules have been followed)
- draw on secondary data you have sourced by yourself

Once the data have been collected, you are required to work entirely independently in order to:

- present, analyse and summarise your findings and data
- draw conclusions and make an evaluation of the completed work

Geographical skills

You are expected to develop various skills as a geographer. Skills are both **quantitative** (using mathematical, computational and statistical procedures to record phenomena and processes) and **qualitative** (using non-numerical techniques such as cartographic and GIS data, visual images, interviewing and oral histories). The specification provides a full list. Some statistical skills appear in the companion volumes to this book.

Specialised concepts

The following terms are essential for a twenty-first century A-level geographer to know and understand. Use them correctly in context whenever you can because the examination questions will expect you to understand what they mean.

Adaptation: The ability to respond to changing events and to reduce current and future vulnerability to change. Societies must adapt to climate change, for instance.

Causality: The relationship between cause and effect. Everything has a cause or causes. For instance, precipitation is an effect of the collision or fusion of smaller water droplets, which in turn have been caused by air uplift and cooling (see p. 28).

Equilibrium: A state of balance between inputs and outputs in a system such as the water of the carbon cycle. A steady-state equilibrium means there is balance in the long term but the system fluctuates in the short term.

Feedback: The way that environmental changes become accelerated, or are negated, by the processes operating in a human or physical system (see p. 63).

Globalisation: The process by which the world is becoming increasingly interconnected as a result of increased integration and interdependence of the global economy. Globalisation may be viewed as a cause of accelerated anthropogenic (human) climate change.

Identity: How people view changing places, landscapes or societies from different perspectives and experiences. Climate change is predicted to bring significant changes to landscape identity in the future (such as dominant vegetation species).

Inequality: Social and economic (income and wealth) inequalities between people and places. These inequalities give rise to movements of people. In the future, climate change may exacerbate inequalities.

Interdependence: Relations of mutual dependence — some elements of the water and carbon cycles are interdependent (see p. 62).

Mitigation: The reduction of a phenomenon that is having a negative effect on people, places or the environment. Climate change mitigation is an attempt to reduce greenhouse gas emissions (by an individual or a society).

Place: A unique portion of geographic space. Places can be identified at a variety of scales, from local territories or locations to the national or state level. Places can be compared according to their physical characteristics. For example, the UK's peatlands (p. 52) are highly distinctive places.

Power: The ability to influence and affect change at different scales. Power is vested in citizens, governments, institutions and other stakeholders. Successful climate change mitigation will depend on the attitudes and actions of powerful countries such as the USA and China.

Representation: How a country, place or area is portrayed by formal agencies (government and businesses) and informally by citizens. The UK's peatlands are represented increasingly as carbon sinks (p. 56) in order to justify their conservation.

Resilience: The ability of an environment or society to adapt to changes that have a negative impact upon them, such as water shortages.

Risk: The possibility of a negative outcome resulting from a physical process, such as a storm or drought.

Scale: Places can be identified at a variety of geographic scales, from local territories to the national or state level. Global-scale interactions occur at a planetary level.

Sustainability: Development that meets the needs of the present without compromising the ability of future generations to meet their own needs.

System: A set of interrelated objects. A system can be either closed — with no import or export of materials or energy across its boundary (e.g. the global water cycle) — or open, where imports occur (e.g. the local drainage basin water system).

Threshold: A critical limit or level that must not be crossed in order to prevent a system from undergoing accelerated and potentially irreversible change (see p. 64).

Content Guidance

Water and carbon cycles

■ The concepts of system and mass balance

The global water cycle

At the global scale, there is a fixed amount of water in the Earth–atmosphere system, amounting to about 1,385 million cubic kilometres in volume. At any given moment, this water is either held in one of several **stores** or is being transferred between them via a series of **flows** operating over varying timescales.

■ Global water stores include the atmosphere and the ocean; there are numerous terrestrial stores too, including the soil, rivers, lakes, reservoirs and vegetation (Table 1).

■ A water flow entering a store is called an **input**; flows leaving stores are called **outputs**. The global hydrological cycle as a whole is a **closed system** (Figure 1). This means it does not have any external inputs or outputs operating across the system boundary. The system's **mass balance** does not change over time.

■ Water can change state over time and may be stored in vapour, liquid or solid forms. The proportion of global water that is stored in solid form as ice is called the **cryosphere** (Figure 2).

At a global scale, mass balance means that the total amount of water is always conserved (although changes can occur in where it is stored or accumulates). At a local (drainage basin) scale, water system inputs are equal to outputs, plus or minus any changes in storage (or accumulation).

As Figure 1 shows, water moves between stores in a series of transfers, some of which involve a change of state, while others are merely a movement of water from one place to another. For instance, water vapour in the atmosphere is precipitated onto the land as rain or snow after **condensation** occurs. Once on the land, water may evaporate back to the atmosphere from the ground surface itself. Any remaining water sinks into and beneath the ground surface under gravity, and moves slowly towards the sea through the soil or underlying bedrock. Under certain conditions, water that cannot soak into the ground will flow more quickly over the land towards a river, a lake or the sea.

The main global water stores are shown in Table 1. Water is continuously cycled through these stores by flows, which are explored in depth on pp. 17–18.

A **store** is a reservoir where water is held, such as the ocean or ice caps.

A **flow** is a movement (or transfer or flux) between stores in a system.

The **cryosphere** consists of those areas of the Earth where water is frozen into snow or ice, including ice sheets, ice caps, alpine glaciers, sea ice and permafrost.

Table 1 The approximate sizes, characteristics and distribution of the major global water stores

Store	Volume (cubic km)	% of total water	% of fresh water	Distribution and characteristics
Oceans	1,335,040	96.9	0	Oceans cover two-thirds of the planet
Cryosphere	26,350	1.9	68.7	Located at high latitudes (polar regions) and high altitudes
Groundwater	15,300	1.1	30.1	Deep groundwater can remain stored for 10,000 years
River and lakes	178	0.01	1.2	Uneven distribution because of climatic variations
Soil moisture	122	0.01	0.05	Remains permanently or seasonally frozen in northern permafrost regions
Atmosphere	13	<0.01	0.04	There is sufficient moisture for ten days of rain in the atmosphere
Biosphere (vegetation and fauna)	0.6	<0.01	<0.01	Distributed unevenly because of climatic variations

① In the oceans the vast majority of water is stored in liquid form, with only a minute fraction as icebergs.

② In the cryosphere water is largely found in a solid state, with some in liquid form as meltwater and lakes.

③ On land the water is stored in rivers, streams, lakes and groundwater in liquid form. It is often known as blue water, the visible part of the hydrological cycle. Water can also be stored in vegetation after interception or beneath the surface in the soil. Water stored in the soil and vegetation is often known as green water, the invisible part of the hydrological cycle.

④ Water largely exists as vapour in the atmosphere, with the carrying capacity directly linked to temperature. Clouds can contain minute droplets of liquid water or, at a high altitude, ice crystals, both of which are a precursor to rain.

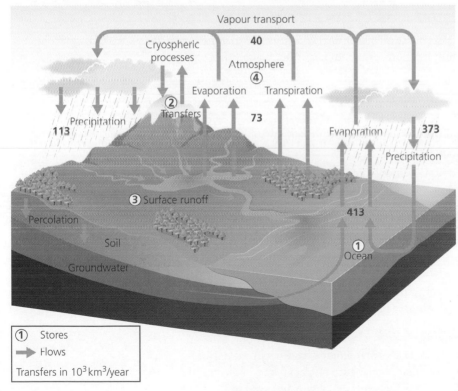

Figure 1 The global water cycle

Knowledge check 1

The relative importance of the water flows and storage shown in Figure 1 varies for different countries and continents. How can climatic factors help explain this?

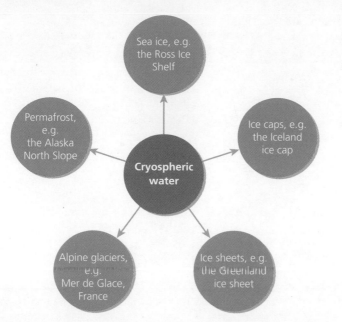

Figure 2 Different locations of water in the cryosphere store

Temporal and spatial changes in the size of water stores

Global hydrological stores vary in size naturally over different timescales. This can be illustrated by briefly examining short-term and long-term changes in cryosphere storage.

Seasonal changes in cryosphere storage

Seasonal variations in levels of ice **accumulation** and **ablation** bring marked cyclical fluctuations in water flows and storage in mountainous glaciated regions such as the Alps, the Andes and the Tibetan plateau.

■ The balance between ablation and accumulation, which is maintained normally from year to year, is described as a **steady-state equilibrium**. This means that the system maintains balance when viewed over the longer term, despite any imbalances between inputs and outputs that exist for particular months or times of the year. Similarly, the extent of Arctic sea pack ice is reduced by about 50% during summer in the northern hemisphere, before reforming each winter.

■ However, recent evidence suggests that these and other cryospheric water stores may actually be experiencing a permanent reduction in size because of anthropogenic (human-induced) climate change (see p. 14).

Long-term changes in cryosphere storage

Significant and far longer-lasting changes in water storage have occurred naturally in the distant past, including the arrival and departure of glacial and inter-glacial epochs. There were periods in Earth's past when the planet is believed to have been entirely ice-free. At other times, the majority of land was probably covered with ice.

■ Geologists believe that the Earth may have become a giant 'snowball' during two distinct Cryogenian ice ages, which occurred around 650–750 million years ago. Cryospheric storage would have been much greater than it is today.

Accumulation is the build-up of snow and ice which takes place in the cryosphere.

Ablation is the change from solid ice to liquid or water vapour when temperature rises above 0°C. This wastage of surface snow or ice is achieved by melting and evaporation (the change from ice to vapour is called sublimation).

- In contrast, during much of the Palaeocene and early Eocene geological epochs (about 65–35 million years ago), 'hothouse' conditions meant the poles were probably free of ice caps — and crocodiles lived above the Arctic Circle! Global sea level would have been much higher than today because more than 99% of all water was stored in the oceans in the absence of any cryospheric water.

- These changes occurred because of long-term natural processes, including three cycles affecting the Earth's orbit around the sun, which bring warming and cooling over long periods of time (as incoming levels of solar radiation change). First, every 100,000 years, the Earth's orbit changes from spherical to elliptical, changing the solar input. Second, the Earth's axis is currently tilted at 23.5°, but this changes over a 41,000-year cycle between 22° and 24.5°, also affecting solar input. Third, the Earth's axis wobbles, changing over 22,000 years, bringing further climate change. These orbital cycles are usually termed **Milankovitch cycles**.

- Plate tectonic movements — such as Antarctica's arrival at the south pole and the subsequent formation of the Antarctic ice sheet — have also had an important influence on changing cryosphere storage (on a timescale of many millions of years).

Changing processes and water transfers

Alongside storage changes, there is great natural variability in how water transfers operate within and between land, ocean, atmosphere and cryosphere stores over varying timescales, from minutes to millennia.

Seasonal changes in rainfall

Significant transfers take place over a relatively short period of time on account of meteorological processes associated with the onset of the annual **monsoon** in east Asia.

- A monsoon is an especially wet season (the term is derived from the Arabic word *mausim,* meaning season), triggered by a periodic alternation of wind direction and velocity affecting some tropical regions of the world.

- The strongest monsoon effects of heavy rainfall are felt in India, Pakistan and China. Monsoons are also found in east Africa and Australia.

- In Mangalore in southwest India, more than two-thirds of the year's rainfall falls in just three months each year (Figure 3). This means that water transfers from the atmosphere to the land, and from the land to the sea, are seasonally uneven.

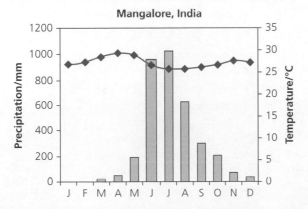

Figure 3 During the east Asian monsoon, a large water transfer occurs between the atmosphere and land during a relatively short time period

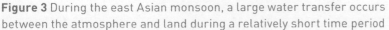

> ### Self-study task 1
>
> Figure 3 shows climatic data for a part of east Asia. Describe the main features of this climate.

Year-to-year climatic and rainfall variability

Many parts of the world experience naturally occurring climatic variability lasting for years or even decades, which can act to reduce or increase water flows in some contexts.

■ The **El Niño southern oscillation (ENSO)** is a natural climatic phenomenon in the Pacific Basin which recurs every 3–7 years and usually lasts for 18 months.

■ During an El Niño event, cool water normally found along the coast of Peru is replaced by warmer water. At the same time, warmer water near Australia and Indonesia becomes cooler.

■ As a result of these and other changes, very dry conditions arrive in parts of southeast Asia, India, eastern Australia, southeastern USA and Central America. In these contexts, precipitation transfers from the atmosphere to the land are greatly reduced for a period of time. In contrast, rainfall increases across the east-central and eastern Pacific Ocean, including the Californian coastline.

Long-term changes in rainfall and aridity

Long-term anthropogenic climate change is projected to have drastic results for water transfer processes around the world. Warmer air can hold more moisture, so global warming is expected to result in large changes in precipitation. Broadly, areas that already have a lot of rain can be expected to experience more, while areas that are already **arid** may suffer lower rainfall. Rainfall patterns are also likely to change, with an increase in very heavy, episodic downpours, perhaps punctuated by longer periods of drought. However, scientists cannot yet predict with much detail the areas that will be most affected. There is great uncertainty here.

■ Areas viewed by scientists as being at high risk of increasing aridity include desert fringes in Australia, China, the USA and the Sahel region (a long strip of semi-arid drylands including parts of Sudan, Chad, Burkina Faso and Niger). Climate data for the Sahel suggest that a long-term reduction in rainfall may be taking place. Rainfall has been lower in recent decades than during earlier decades of the twentieth century.

■ Climate change scientists believe that rainfall patterns are likely to change in the UK as the world's oceans warm, and may in fact be doing so already. The UK may in future become faced with longer, hotter summers (especially in London and the southeast) and warmer, wetter winters, with more rain-bearing storms for all parts of the UK, not just northern regions.

Consider also the water transfers from land to the ocean — and accompanying sea level rise — that would follow if the Greenland and Antarctic ice sheets were to melt. Together, they contain more than 99% of the freshwater ice on Earth (the Antarctic Ice Sheet covers the same area as the USA and Mexico combined; the Greenland Ice Sheet is much smaller). Overall, Antarctica is currently losing slightly more ice than it gains each year. The deficit consists of a water loss of around 70 gigatonnes per year (1 gigatonne is 1 billion tonnes). If the Greenland Ice Sheet melts altogether, scientists estimate that sea level will rise by about 6 m as a result of the transfer of meltwater from land to ocean. If the Antarctic Ice Sheet melts, sea level will rise by a further 60 m.

The **El Niño southern oscillation (ENSO)** is a sustained sea surface temperature anomaly across the central tropical Pacific Ocean. Along with La Niña events, El Niño events are part of a short-term climate cycle lasting 1–2 years.

Arid conditions means a severe lack of water, usually defined as annual rainfall totalling less than 200–250 mm.

Exam tip

Global sea level has risen by 200 mm since 1900. Be sure to explain that this change — attributed to global warming by most scientists — is primarily due to thermal expansion of the oceans (water actually expands slightly in volume as its temperature rises, like liquid in a thermometer) rather than water transfers from melting ice.

Summary

- The global water cycle consists of numerous stores and flows. It is a closed system in which all water (and mass balance) is conserved over time. The flows into and out of different stores are called inputs and outputs.
- In addition to lakes, oceans, the atmosphere, vegetation, soil and groundwater, water is stored in the cryosphere. This ice store is the most important global store of fresh water in terms of its size.
- Historically, the relative size of ocean and cryosphere storage has altered considerably between different geological epochs. In the future, anthropogenic climate change may lead to further changes.
- Water transfers and stores can increase or decrease in size on shorter timescales too, for example the annual monsoon in Asia, or water transfers taking place after annual snowmelt in cold regions.
- El Niño events bring changes in water processes that can last for more than a year

Catchment hydrology: the drainage basin as a system

The drainage basin system

The **drainage basin** is a subset of the global hydrological cycle, and is defined as a catchment area forming part of the Earth's surface area which is drained by a particular stream or river. Drainage basins can vary in size from several square kilometres (a small catchment in England or Wales, for instance) to many thousands of square kilometres in the case of a large continental river such as the Mekong or Colorado.

All land belongs in one drainage basin or another. The dividing line between adjacent basins is called a drainage divide, or **watershed**. For small local-scale rivers, slight variations in relief can help to divide the land surface into different drainage basins. In the case of continental rivers, the drainage divide may correspond with larger-scale relief features, such as a range of mountains. Unlike the global-scale water cycle, the drainage basin is an **open system**. The main inputs, stores, flows and outputs of the drainage basin system are portrayed in Figure 4 in two contrasting ways.

An **open system** allows energy and matter to be transferred across its boundary from external areas.

Drainage basin inputs

Drainage basin inputs consist of different types of precipitation, including rain, snow, sleet, hail and frost. Alongside the type and amount of precipitation, its duration and intensity affect how a drainage basin system responds.

- High intensity rainfall of between 50 and 100 mm per hour is rare in the UK. When it does occur, however, the result can be flash flooding. For instance, the devastating Boscastle floods of 2004 were a result of 185 mm rainfall arriving in just five hours (equivalent to around two months of rainfall normally). This kind of occurrence is rare: the intense rain experienced by Boscastle was a once-in-400-year event. In December 2016, the worst flooding in 50 years hit Maesteg in south Wales.

Exam tip

Drainage basin hydrological inputs and flows must be studied in greater depth than at GCSE, particularly by students who plan to use this topic as the basis for their independent investigation.

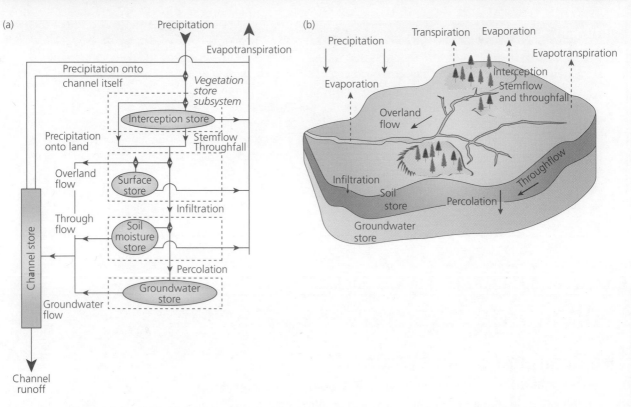

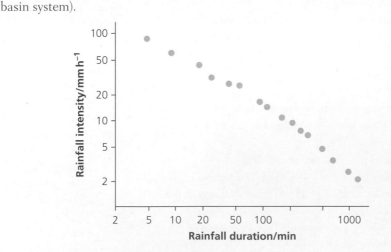

Figure 4 A system diagram and a sketch of the hydrological cycle for a river drainage basin

- Low intensity but relatively long duration rainfall is far more common in the UK.
- Figure 5 shows the relationship between rainfall intensity and duration for a small tributary of the River Thames. As you can see, it is unheard of to have high intensity rainfall lasting for very long, whereas low intensity rainfall often lasts for many hours (Table 4 on p. 31 also shows rainfall events of contrasting intensity and duration, and analyses the effects of differing precipitation inputs on a drainage basin system).

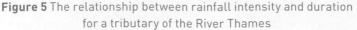

Figure 5 The relationship between rainfall intensity and duration for a tributary of the River Thames

Fieldwork

Precipitation data for a local area can be collected using rain gauges. Samples can be collected to compare how much water reaches the ground at sites with contrasting types of vegetation cover, such as open grassland or woodland.

Drainage basin flows

Precipitation inputs may be transferred through a drainage basin via a number of different flows.

- Rainwater dripping from leaves and branches towards the ground is called **throughfall**. Water that falls directly onto vegetation but flows to the ground via stems and trunks is termed **stemflow**.
- **Infiltration** is the movement of water from the ground surface into the soil. The capacity of the soil to transmit any water that actually falls on it is a crucial valve in the hydrological system. The rate at which water can pass into the soil is known as its **infiltration capacity**, usually expressed as millimetres per hour. Rainwater will be held on the ground surface if it falls at a rate that is greater than the infiltration capacity. Should this occur, it is only a temporary form of storage: some water may begin to flow from surface puddles downhill over the land towards the river or ocean, or it may evaporate directly to the atmosphere. In the case of falling snow, all precipitation is stored temporarily on the surface until melting begins.
- **Throughflow** is the movement of water laterally (sideways) through the soil, via a matrix of pore spaces, fissures and pipes (wide gaps in the soil created by roots and animal burrows). With the exception of pipeflow, throughflow movements are relatively slow (Table 2). Throughflow is most effective in the surface horizons of the soil because these are, in general, less compacted and have high **permeability**. On farmed land, the surface horizon has often been ploughed and so the soil structure is open, with many large spaces and fissures which water can soak into. In contrast, lower soil horizons are more compacted by the weight of overlying material, which reduces the soil's permeability. When this happens, water that is soaking downwards under gravity is deflected laterally down slope in the soil.
- **Percolation** is the transfer of water from the soil into the underlying bedrock.
- **Groundwater flow** is the vertical and lateral movement of water through a drainage basin's underlying rock as a result of gravity and pressure. All rocks

Table 2 Flow estimates for hydrologic processes

Type of flow	Flow route	Velocity (m/h)
Overland flow	■ Over the ground	50–500
Soil throughflow	■ Matrix movement (water flows between soil pores)	0.005–0.3
	■ Pipeflow (animal burrows)	50–500
Groundwater flow	■ Jointed limestone	10–500
	■ Sandstone	0.001–10
	■ Shale	Negligible
The stream	■ Open channel	300–10,000

Exam tip

Remember that precipitation includes rain, snow, sleet, hail and frost. If an exam question asks you to analyse or explain changes over time in precipitation and water flows, remember that snowfall and snowmelt could play a part in the pattern.

Permeability is the ease with which water (or gas) can pass through rocks or a soil horizon.

Knowledge check 3

Study Table 2, which shows differences in the rate of flow for the different hydrological processes. Can you suggest reasons for these differences?

show some **porosity** (the volume of voids as a percentage of the bulk volume of material) and resulting permeability. However, porosity and permeability vary enormously according to rock type. Coarse-grained sedimentary rocks show high permeability and permit high levels of groundwater flow, whereas fine-grained igneous rocks are relatively impermeable. Below a certain depth, the bedrock of a drainage basin may be permanently saturated. This level below which the ground is saturated is called the surface of the **water table** and it can vary in depth according to the season.

The importance of overland flow

In addition to movements into and through the soil and rock, water can sometimes flow *over* the surface at very fast rates. **Overland flow** (also called **surface runoff**) is the movement of a sheet of water across the ground surface towards a river, lake or ocean, sometimes at a very fast speed. Overland flow can occur after either long duration or intense rainfall.

1 **Saturation excess overland flow** happens if rainfall continues for a long time. All soil layers become saturated and throughflow is deflected closer and closer to the surface. This is because over time during a rainfall event the upper and more permeable soil horizons become saturated as the water level in the soil rises (exactly the same thing happens if you pour too much water into a plant pot: the level of standing water rises until it reaches the surface). In time the entire soil becomes saturated right up to the surface. Thereafter, saturation-excess overland flow begins.

2 **Infiltration-excess overland flow** is defined as overland flow which occurs when rainfall intensity is so great that not all water can infiltrate, irrespective of how dry or wet the soil was prior to the rainfall event. This process, first described by R.E. Horton in 1945, is extremely common in some parts of the world, especially semi-arid areas where high intensity rainfall encounters hard-baked ground with a relatively low infiltration capacity. In these environments, infiltration-excess overland flow can lead to flash flooding and the growth of large, deep channels called *wadis* which fill with water when it rains. In contrast, the process is less common in parts of the world with a humid temperate climate, such as the UK. You may, however, have witnessed a rare infiltration-excess overland flow event following a torrential downpour of summer rain, which falls so fast that a sheet of water begins immediately to flow across the surface of the land: there is simply no time for the water to soak into the ground or run into drains.

Figure 6 shows the operation of both types of overland flow.

An important aspect of overland flow dynamics is the way that additional parts of a drainage basin begin to contribute to saturation-excess overland flow during a long-lasting storm event. As we have seen, under normal (i.e. non-intense) rainfall conditions, downslope overland flow will only occur once the ground below is completely saturated and water can no longer soak in. Areas close to the bottom of a slope tend to become saturated first during a storm because they are receiving throughflow from higher up the slope in addition to directly infiltrating rainfall. As time continues, this saturated area begins to extend further up the slope towards the crest of the hill (Figure 7). The saturated zone that develops is called the **saturated wedge** because it has a triangular or wedge shape upslope.

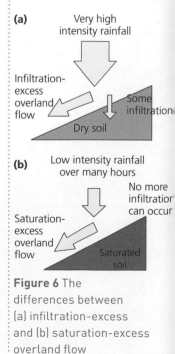

Figure 6 The differences between (a) infiltration-excess and (b) saturation-excess overland flow

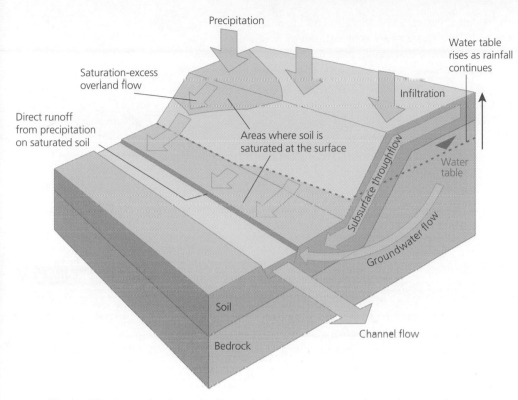

Figure 7 Drainage basin water flows during a storm event (note the growth of a saturated zone at the base of the slope)

Drainage basin stores

Several drainage basin stores are also shown in Figure 4. They include the:

- **interception store** (leaf and plant surfaces)
- **vegetation store** (water held in the biomass itself, also called 'green water')
- **surface store** (water collected on the surface of the ground in depressions and hollows, and also snow cover)
- **soil moisture store** (water held in pore spaces in the soil matrix)
- **channel store** (water held in the river channel itself at any moment in time)
- **groundwater store** (water stored in solid rock and in any superficial deposits, e.g. gravels below the soil)

In some local contexts, **interception** may prevent almost all of the precipitation falling on an area from ever reaching the ground surface. Interception varies with the duration of precipitation and with the character of the vegetation. If a storm lasts a short time then a considerable proportion of the rainfall remains caught on the leaves and branches. All of this water may be evaporated eventually back to the atmosphere. However, during a longer storm these myriad small reservoirs will overflow and the rainfall drips to the surface as throughfall, or flows along vegetation surfaces to the ground as stemflow. As the duration of the storm increases so too does the proportion of rainfall reaching the surface beneath the plants.

Exam tip

If a question asks you to discuss how water stores and flows can change over time, think about the changes that can take place in just minutes or even seconds during a storm event. The zones of saturation shown in Figure 7 can change in size very quickly.

Interception is the temporary storage of precipitation on the leaves or branches of a plant.

Of course, vegetation and land use may vary considerably from place to place in a drainage basin. As a result, interception storage is often uneven. In addition to any spatial variations, there can be marked seasonal differences in interception for drainage basins in climate zones with deciduous vegetation that sheds its leaves in autumn. Deciduous woodland in England and Wales has a greatly reduced ability to store water during winter months. Looking further afield, the capacity for interception storage of different global **biomes** varies enormously.

- Tropical rainforest trees are well adapted to a hot humid equatorial climate by possessing 'drip tip' leaves: these encourage a more rapid flow of rainwater through the tree canopy and towards the ground via throughfall and stemflow.
- High-latitude coniferous trees have sloping branches: snow slides towards the ground instead of building up on the branches, which might cause them to break. As a result of this plant adaptation, snow is transmitted from the vegetation store to the drainage basin's surface store.

The effectiveness of surface or depression storage depends upon local relief factors. If the drainage basin contains relatively few flat areas or depressions and hollows, then there is little potential for depression storage. However, if the land is relatively flat and contains hollows where water can collect, then a significant amount of water can be stored on the surface, especially if the ground is already saturated or has a naturally low infiltration capacity. The collection of large amounts of surface water can result in a condition known as **pluvial flooding** for urban areas.

The soil moisture store

There are actually three different types of soil water:

1 Water adhering in thin films by molecular attraction to the surface of soil particles is called **hygroscopic water** (this form of water is not available for plants). Overall, this is a relatively insignificant form of water storage.
2 Water forming thicker films and occupying the smaller pore spaces in the soil is termed **capillary water**. It is held against the force of gravity by surface tension, and is available for plants to use. This is the water that remains in the soil when excess water has drained away after a storm event. It is vitally important for plant health.
3 The excess water that occupies all large and usually free-draining spaces in the soil is called **gravitational water**. This transitory water drains away soon after rain stops falling. If it did remain, the soil would become permanently waterlogged and unable to support much vegetation.

Figure 8 shows different states of soil moisture storage, including **field capacity** (the total amount of water remaining in a freely drained soil after all gravity water has been drained away following the end of rainfall) and **wilting point** (when there is insufficient soil water to compensate for plant water losses from transpiration).

Exam tip

In addition to being able to describe different water stores, you may need to discuss interconnections between them. For instance, the vegetation store influences the characteristics of the soil moisture store. Plant roots help to aerate the soil; humus derived from organic matter helps retain soil water.

A **biome** is a plant community whose global distribution corresponds with a climatic region of the Earth, for example the tropical rainforest or Arctic tundra.

Knowledge check 4

What are the names of the main global biomes? How and why might levels of interception storage vary for different types of global biome?

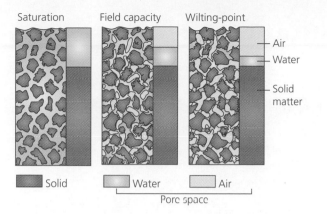

Figure 8 Different states of soil water storage and proportions of air, water and solid matter

Drainage basin outputs

There are three principal catchment outputs.

1 **Evaporation** is the change in state of water from a liquid to a gas. For this to happen, heat energy is required. Evaporation can occur from the surface of any water store including the ocean, surface water on the land and water intercepted temporarily on plant leaves. Many meteorological factors influence the rate of evaporation, including temperature, humidity and wind speed.

2 **Transpiration** is the diffusion of water from vegetation into the atmosphere, involving a change from liquid to gas. Water is lost through the stomata (pores) of leaves, and different types of vegetation vary greatly in terms of the rates of transpiration that are allowed. Tropical trees have large leaves, which maximises their rates of transpiration, while in contrast coniferous trees have needle leaves which minimise transpiration. The term **evapotranspiration** is used to describe a combination of evaporation and transpiration.

3 **Channel discharge** is the volume of water leaving a drainage basin via its main stream or river during a specified unit of time. A river's discharge is generally measured in cumecs (cubic metres per second).

Temporal and spatial variations can be studied and observed in the relative importance of the three system outputs. Marked seasonal changes can be observed in many local and regional contexts, while different types of vegetation are adapted to the environment in ways that may maximise or minimise their rates of transpiration.

Summary

- Water flows and stores can be studied at the local scale by examining the characteristics and functioning of individual river drainage basins. Variations in precipitation type, amount, duration and intensity all affect how the river may respond to inputs into its drainage basin.
- The pattern of flows through the drainage basin can be complex and involve many different processes. These include different types of overland flow, in addition to stemflow, infiltration, throughflow, percolation, groundwater flow and channel flow.
- The interception store plays an important role in preventing or allowing rainwater to move to other stores in the drainage basin system.
- Local variations in soil and rock type play an important role in catchment hydrology.
- Water leaves the drainage basin as one of three outputs: evaporation, transpiration or channel discharge.

Temporal variations in river discharge

River regimes and the factors influencing them

A **river regime** can be defined as the annual variations in the pattern of flow, or discharge, of a river, measured at a particular point (such as a gauging station). Some of the annual river flow is supplied by **storm flow**: overland flow and throughflow following consecutive rain or snow events. Much is also supplied from groundwater between periods of precipitation. This feeds steadily into the river, and gives rise to a normal minimum flow of the river, which is called its **baseflow**.

The character of a river regime is influenced by several physical and human factors:
- Physical factors include the annual pattern of precipitation, snowmelt, temperatures and evaporation, along with the factors of relief, vegetation and the underlying soil and geology.
- Human factors include the construction of dams and reservoirs, levels of irrigation and the extent to which water is removed from a river through transfer schemes that help supply other places with water.

Some rivers have relatively **simple regimes**: periods of high and low channel flow may correspond with seasonal temperature and rainfall changes, perhaps incorporating a monsoon or spring snowmelt event. In contrast, some of the largest continental rivers have relatively **complex regimes**, including the Colorado River in its natural state prior to the construction of the Glen Canyon Dam in 1966 and Hoover Dam in 1935.
- Figure 9 shows that the Colorado's regime used to include several marked peaks and troughs spread throughout the year.
- Originally, the river had a mostly very high flow between April and September. During the summer months, snowmelt in the Rocky Mountains and Winter River Mountains caused a significant rise in water level along the entire river. In the most extreme years of the early twentieth century, the Colorado's discharge was thirteen times higher in midsummer than in winter.

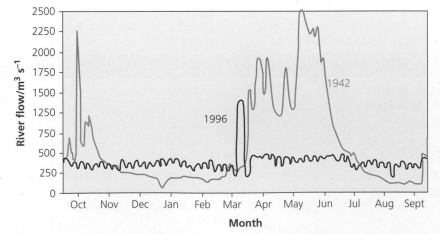

Figure 9 River regime of the Colorado at Lee's Ferry, Arizona, in 1942 and 1996

Knowledge check 5

How has the relative importance changed over time for different physical and human factors affecting the Colorado River's regime?

- However, the peak flows recorded from April onwards consisted of a number of separate spikes, each corresponding with snowmelt arriving from a different tributary river somewhere in the Colorado's enormous drainage basin (which covers seven US states). Mountain snowpack begins to melt in different upper regions of the catchment at slightly different times.
- Although river flow used to be generally low from September onwards, violent autumn storms over the Colorado Plateau contributed to another high, although shorter lived, peak discharge in the autumn months.
- The construction of large dams across the Colorado River has smoothed out the naturally occurring flood peaks and periods of low flows, as Figure 9 shows (the brief discharge 'spike' in March 1996 was caused by the deliberate release of water from the dam). The Hoover Dam now stores the equivalent of two years' river flow in Lake Mead. The mean annual flood at the gauging station at Lee's Ferry has been reduced from 2264 cubic metres per second to 764 cubic metres per second (according to data collected in the 1980s).

Figure 10 shows two contrasting river regimes for small rivers in the UK. Climate, season, geology, vegetation and land use have all influenced the patterns shown.

- River A shows lower flows overall, reflecting lower precipitation in the southeast of England.
- Seasonal variations in precipitation and river flow are observable in both cases.
- Differences in geology, vegetation and land use help to explain the time differences between periods of high rainfall and peak river flows. River A's drainage basin is an area of chalk grassland: high levels of interception, in combination with permeable geology, result in the slow and steady transfer of water via infiltration, percolation and groundwater flow (hence the peak river flow period is recorded several months after winter rainfall has peaked). In contrast, River B's drainage basin is an upland area of overgrazed land in Scotland, underlain by impermeable granite. Intense winter rainfall runs quickly over the land towards the river, resulting in a much closer correlation between annual peaks in rainfall and river flow. You can also observe the results of heavy snowmelt in April.

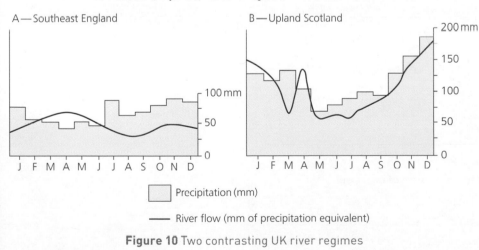

Figure 10 Two contrasting UK river regimes

Self-study task 2

Explain how climatic and geological factors may have affected the rainfall and river flow patterns shown in Figure 10.

Storm hydrograph shapes and components

A storm hydrograph shows a river's response to a specific input of precipitation. The main features of a typical hydrograph are labelled in Figure 11:

- **peak discharge:** the maximum rate of flow during a storm event
- **peak rainfall:** the maximum rainfall recorded in one of the time intervals
- **rising limb:** the part of a storm hydrograph in which the discharge starts to rise
- **lag time:** the time elapsed between peak rainfall and peak discharge
- **falling limb:** the part of the storm hydrograph in which the discharge starts to fall
- **preceding discharge:** the rate of flow prior to the latest storm event (if there has been little rainfall previously, this will be the river's normal baseflow level)
- **bankfull discharge:** the maximum discharge reached during a storm event prior to overtopping of the river banks and the inundation of the floodplain with excess water
- **baseflow:** the normal minimum flow of the river (supplied by groundwater flow)

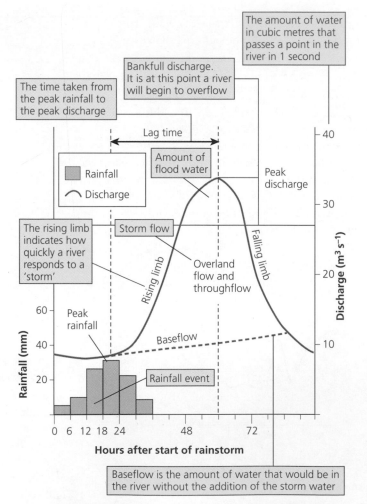

Figure 11 The features of a storm hydrograph

Climatic influences on storm hydrographs

As we have seen, rainfall intensity and duration have an important influence on the operation of overland flow, and it is this more than any other climatic influence that determines whether the river has a 'flashy' response (i.e. a high peak discharge and low lag time). **Antecedent conditions** also play a large role. If the basin's soil was already saturated because of previous rainfall, then overland flow would have occurred early on in the storm event, resulting in a flashy hydrograph. In contrast, low intensity rainfall on relatively dry ground is unlikely to produce a flashy hydrograph (nor will peak discharge be especially high, unless the rainfall is of extremely long duration).

The seasonal effects of evaporation (determined by temperature) and transpiration rates are also important. The flashiest UK hydrographs are often associated with high intensity winter rainfall falling on previously saturated soil at a time of year when evaporation and transpiration outputs from the drainage basin are minimal.

The type of precipitation matters too: snow can be stored on the ground for months before melting, resulting in a greatly delayed peak discharge.

Non-climatic influences on storm hydrographs

Other river catchment characteristics influence storm hydrographs (Figure 12). We have already explored how factors such as vegetation and soil type may influence the operation of catchment flows such as throughflow, groundwater flow and overland flow. As a general rule:

- steep relief and impermeable geology favour flashy hydrographs, whereas free-draining sandy soils or highly permeable or porous geology can lower an area's overland flow and flood risk
- flashy hydrographs are more likely to be seen in poorly drained upland regions of the UK where grass or peat moorland dominates, resulting in limited interception storage (and low evaporation rates)

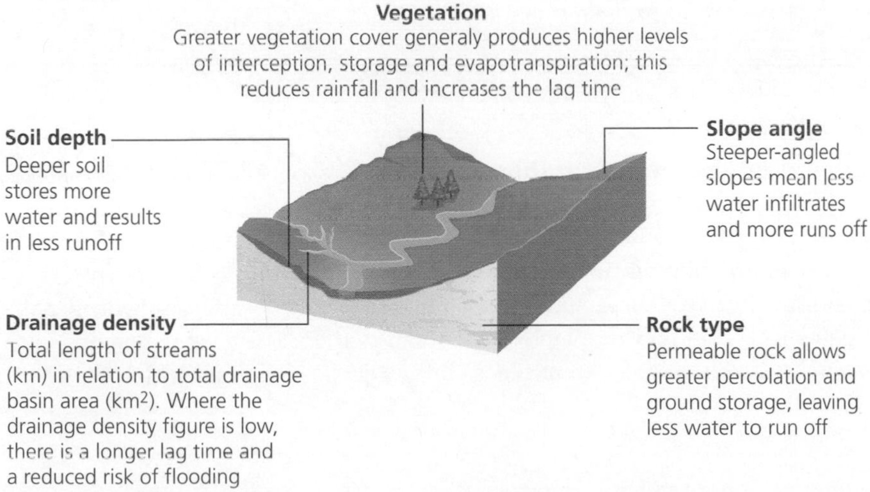

The weather and soil moisture conditions immediately prior to a storm event are called **antecedent conditions**.

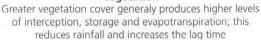

Vegetation
Greater vegetation cover generaly produces higher levels of interception, storage and evapotranspiration; this reduces rainfall and increases the lag time

Soil depth
Deeper soil stores more water and results in less runoff

Slope angle
Steeper-angled slopes mean less water infiltrates and more runs off

Drainage density
Total length of streams (km) in relation to total drainage basin area (km²). Where the drainage density figure is low, there is a longer lag time and a reduced risk of flooding

Rock type
Permeable rock allows greater percolation and ground storage, leaving less water to run off

Figure 12 How physical factors such as rock type, slope angle and vegetation affect runoff and discharge levels

Soils and land use

At the local scale, soil type variations affect how much infiltration and throughflow can occur. Some drainage basins have a clay soil type with a low infiltration capacity (see Table 3). Agriculture and land use practices also have a significant impact on hydrographs at the local level. Well-established pasture has a relatively high infiltration rate: tall grasses can store water on their surfaces which is transmitted steadily via stemflow to the soil below, whose infiltration capacity will be high because it is well aerated as a result of dense grass root networks. Garden lawns in urban areas operate as an effective infiltration valve for the same reason. The replacement of lawns with paved areas and parking spaces is one reason why hydrograph peaks have increased for many urbanised drainage basins in recent decades.

Table 3 Infiltration rates vary according to land use and soil types, with implications for storm hydrograph shapes

Land use	Infiltration rate (mm/h)
Well-established pasture	57
Heavily grazed or 'poached' pasture	13
Cereal	10
Bare compacted soil (average for all soil types)	6

Soil type (bare ground)	Infiltration rate (mm/h)
Clays	0–4
Silts	2–8
Sands	3–12

Fieldwork

This topic lends itself well to an A-level independent investigation. There are many simple fieldwork techniques that can be used to study vegetation density, infiltration rates, and soil texture. For example: how does land use affect hydrological flows?

Self-study task 3

Describe and suggest possible reasons for the variations in infiltration rates shown in Table 3.

Basin size, shape and drainage density

Finally, the size and shape of a drainage basin, along with its **drainage density** (a measure of the length of the river channel and its tributaries per unit area), will influence how a river responds to a rainfall event. This is because the drainage basin shape and channel network density play an important role in determining how quickly water from different parts of the catchment converge at a gauging station on the lower course of the main river channel. Figures 13 and 14 provide simple illustrations of the different hydrograph responses that can be expected for different types of drainage basin shape and drainage density.

Self-study task 4

Describe and suggest possible reasons for the different hydrograph shapes shown in Figure 13.

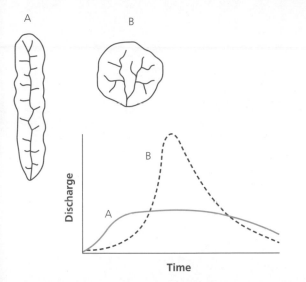

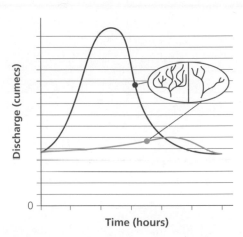

Figure 13 The influence of drainage basin shape on flood hydrographs (circular basins often have flashier flows)

Figure 14 The influence of drainage density on flood hydrographs (basins with high drainage density have flashier flows)

Summary

- A river regime graph shows how the river responds to variations in precipitation over the course of the year. Large continental rivers, such as the Colorado, have naturally complex regimes because of the number of different climate zones, including mountainous regions, found in their drainage basins. Geology and vegetation may also play a role.
- Human factors can alter regime characteristics drastically, particularly dam and reservoir construction, and other land use changes.

- The key features of storm hydrographs include: peak discharge, lag time, rising limb and falling limb.
- Hydrograph shapes vary on account of climatic factors, particularly precipitation duration and intensity, along with seasonal variations in temperature and evaporation.
- Storm hydrographs may also reflect drainage basin characteristics, particularly river catchment shape, drainage density and local influences, which include soil and rock type, slope angle, vegetation and land use.

■ Precipitation and excess runoff in the water cycle

For the water cycle to function, water vapour in the atmosphere needs to change state into precipitation and fall to the ground. For this to happen, **condensation** must take place first. It is achieved either by the cooling of air below its dew point, or by continued evaporation resulting in saturation of the air. Condensation leads to cloud formation and possibly precipitation. Water droplets in the atmosphere form typically when condensation takes place around small dust or sea salt particles (these tiny solids are called **condensation nuclei**).

Condensation is the process by which vapour changes into a liquid or solid form. For this to happen in the atmosphere, condensation nuclei must also be present.

Air uplift and condensation

For precipitation and excess runoff to occur in the water cycle, certain conditions need to be met. First, air must rise in the atmosphere and cool to its saturation point while doing so.

■ The reason why air uplift leads to cooling and condensation is because of the fall in pressure with altitude.

■ This results in the expansion of air (you can observe this in the way balloons increase in size when they rise up into the atmosphere).

■ As a result, there are fewer collisions between air molecules. This reduces the amount of heat energy per unit volume and leads to a fall in air temperature.

■ Cloud formation take place once air temperatures have fallen low enough for water vapour to condense into water droplets.

Reasons for air uplift

There are three main triggers for the uplift of cooling air (Figure 15).

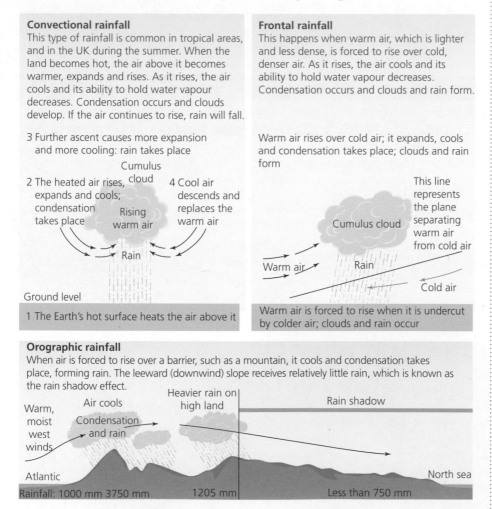

Convectional rainfall
This type of rainfall is common in tropical areas, and in the UK during the summer. When the land becomes hot, the air above it becomes warmer, expands and rises. As it rises, the air cools and its ability to hold water vapour decreases. Condensation occurs and clouds develop. If the air continues to rise, rain will fall.

3 Further ascent causes more expansion and more cooling: rain takes place

Cumulus cloud

2 The heated air rises, expands and cools; condensation takes place
Rising warm air

4 Cool air descends and replaces the warm air

Rain

Ground level

1 The Earth's hot surface heats the air above it

Frontal rainfall
This happens when warm air, which is lighter and less dense, is forced to rise over cold, denser air. As it rises, the air cools and its ability to hold water vapour decreases. Condensation occurs and clouds and rain form.

Warm air rises over cold air; it expands, cools and condensation takes place; clouds and rain form

This line represents the plane separating warm air from cold air

Cumulus cloud

Warm air → Rain

Cold air

Warm air is forced to rise when it is undercut by colder air; clouds and rain occur

Orographic rainfall
When air is forced to rise over a barrier, such as a mountain, it cools and condensation takes place, forming rain. The leeward (downwind) slope receives relatively little rain, which is known as the rain shadow effect.

Warm, moist west winds

Air cools

Condensation and rain

Heavier rain on high land

Rain shadow

Atlantic

North sea

Rainfall: 1000 mm 3750 mm   1205 mm   Less than 750 mm

Figure 15 The three types of rainfall

1 **Convectional rain** results from intense daytime heating of the land. Air parcels adjacent to the ground heat up by conduction; they rise, cool and may form cumulus or cumulonimbus cloud.

2 **Frontal rain** forms when two surface air streams meet. For instance, when polar (cold) and tropical (warm) air masses meet over the North Atlantic ocean in the mid-latitudes, the latter will rise up over the former under low pressure conditions (which result from divergent air flow in the upper atmosphere).

3 **Orographic rainfall** is the result of uplift related to relief features, augmented by the so-called **feeder–seeder mechanism**. The effects are most obvious when warm maritime air encounters a mountainous coastal margin (for example, along the west coast of Scotland or the Cumbrian coastline). As a consequence, the area beyond a relief barrier may suffer from a 'rain shadow' effect, with much drier conditions.

Theories of precipitation formation

Air uplift and condensation does not automatically lead to precipitation. Not all clouds produce rain. An average rain droplet is 1 million times larger than a cloud droplet. In order for rain to fall, cloud droplets have to undergo a rapid process of growth or fusion. The two recognised theories of rainfall production are (1) the **Bergeron-Findeisen process** of ice-crystal growth and (2) the **collision process**, wherein 'super' condensation nuclei generate large heavy water droplets which collide with smaller droplets, sweeping them along into their wake.

The Bergeron-Findeisen process

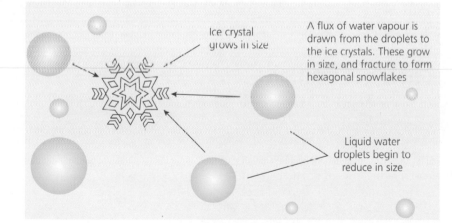

Ice crystal grows in size

A flux of water vapour is drawn from the droplets to the ice crystals. These grow in size, and fracture to form hexagonal snowflakes

Liquid water droplets begin to reduce in size

Figure 16 The Bergeron-Findeisen theory of ice crystal growth and rainfall production

This widely accepted theory of precipitation formation was first developed in the 1930s.

■ Clouds at high altitude — with temperatures just below 0°C — contain a mixture of water droplets and ice crystals.

■ The ice crystals grow quickly at the expense of the water droplets (because of a rapid flux of water vapour from the droplets towards the ice crystals, which happens for relatively complicated scientific reasons).

The **feeder–seeder mechanism** is a process that increases levels of orographic rainfall. Water droplets or ice particles from high altitude 'seeder' clouds fall through a lower-level orographic stratus cloud (the 'feeder' cloud), collecting more cloud water as they do so, which results in heavier rainfall.

Exam tip

When writing about the influence of different uplift mechanisms, try to progress beyond merely explaining each one in isolation. Additionally, you might consider how the three mechanisms could interact, for instance when a weather front crosses a relief barrier.

- The ice crystals fracture as they are jostled by fast high-altitude air currents. Hexagonal ice shapes begin to develop with an even larger surface area. Now even more vapour can condense around these larger snowflake shapes, accelerating their rate of growth even further.
- Eventually, the hexagonal snowflakes become too large and dense to be held aloft. In falling to the ground, they pass through warmer air and melt to produce rain.

In support of this theory, precipitation falling over the UK is often derived from clouds with temperatures below −5°C. Also, the artificial practice of 'cloud seeding' with 'dry ice' (frozen crystals of carbon dioxide) can generate rainfall successfully in accordance with the Bergeron-Findeisen process. However, the theory cannot explain the formation of rain in the warm tropics, where cloud temperatures are higher.

The collision process

An alternative explanation exists that particularly helps to explain rainfall formation in the warm tropics.

- According to the collision theory, 'super-sized' condensation nuclei, such as large sea salt particles, provide 'seeds' around which water droplets form. These are far larger and heavier than normal-sized droplets.
- The larger 'super' droplets fall and collide with smaller droplets by sweeping them into their wake and absorbing them.
- The theory is associated with experimental work carried out by Langmuir, who argued that the higher terminal velocity of large droplets allows them to overtake and absorb many smaller droplets, thereby causing rapid fusion and raindrop growth to occur.

In support of this theory, unexpected downpours and flash flooding in arid areas demonstrate how high numbers of large raindrops can be generated quickly.

Causes of excess runoff generation

Excess runoff generation via overland flow or throughflow following precipitation may occur naturally due to physical factors, including:

- snowmelt and ice ablation in glaciated areas, resulting in seasonal variations in runoff for rivers fed by glaciers, such as the Colorado (see p. 22)
- storm activity, which brings prolonged or high intensity precipitation leading to saturation-excess overland flow and infiltration-excess overland flow, respectively. In the UK, the causes include prolonged frontal precipitation and intense convectional thunderstorms in summer (see p. 28). Table 4 shows hydrological data for a small river basin in Wales which has been affected by storms of varying intensity and duration throughout the year. As you can see, a high percentage of runoff has been generated by some but not all rainfall events
- the monsoon, which brings torrential rainfall and widespread flooding every year to parts of southeast Asia (see p. 13)

Excess runoff (overland flow) generation may also have human causes. Changes in land use can either increase or decrease overland flow and the risk of flooding in a drainage basin. In particular, hydrographs tend to become flashier as a result of (1) catchment urbanisation and (2) widespread deforestation.

Knowledge check 6

Make sure you can explain the Bergeron-Findeisen and collision processes in a clear and error-free way.

Exam tip

'Runoff' is sometimes used to describe all of the water leaving the land via overland flow and throughflow.

Self-study task 5

Using data from Table 4, try to explain the variations in peak discharge for each of the four storm events.

Table 4 Analysing how precipitation duration and intensity affects runoff generation and peak discharge

Storm	Rainfall data			Discharge data		Runoff data	
	Storm amount (mm)	Average intensity (mm/h)	Maximum intensity (mm/h)	Preceding discharge (litres/second)	Peak discharge (litres/second)	Average catchment runoff (mm)	% of precipitation output as runoff
A	12	3	10	50	1,030	5	42
B	10	2	3	70	690	3	33
C	30	3	4	10	1,020	4	13
D	16	1	3	80	660	6	37

Urbanisation and overland flow generation

Figure 17 shows (1) a sequence of changes associated with the progressive urbanisation of a drainage basin, and (2) increases in the size and frequency of high river discharges as urban development takes place in a catchment.

- Urbanisation renders previously permeable ground surfaces impermeable.
- Surfaces such as concrete and tarmac increase surface runoff generation and decrease the effectiveness of infiltration, throughflow and soil storage.
- The more ground that is covered by impermeable hard surfaces such as concrete or paving slabs, the less rainfall will soak into the ground and the more will run over the surface into drains and sewers.
- The drain and sewer network is effectively an extension of the river channel network which allows water to be drained from the land even faster.
- There will be less transpiration and interception without vegetation, which means yet more runoff can be expected.

Self-study task 6

Using information from Figure 17, describe and explain how urbanisation affects the magnitude and frequency of flood discharges.

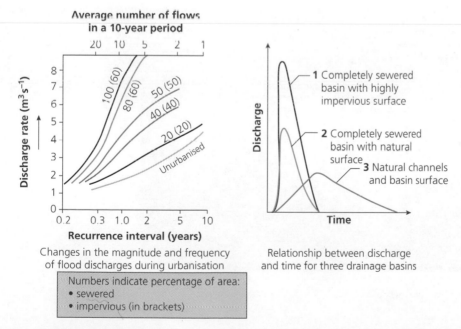

Changes in the magnitude and frequency of flood discharges during urbanisation

Relationship between discharge and time for three drainage basins

Numbers indicate percentage of area:
- sewered
- impervious (in brackets)

Figure 17 The impacts of the spread of urbanisation on runoff and river discharge

Deforestation and overland flow generation

In England, Wales and Scotland, the forest that once covered much of the land has been removed over centuries. Forest land coverage reached an all-time low of 3% in the UK in 1919. Currently, it stands at just over 10% (a large proportion of this recovery can be attributed to coniferous plantations established by the Forestry Commission).

■ Flood risk in some parts of the UK is undoubtedly higher as a result of this changing land use over time.

■ One study in mid-Wales found that rainwater's infiltration rate into the soil was 67 times higher under trees than under sheep pasture.

■ Devastating floods affecting Cumbria in 2009 and 2015 were caused by excess overland flow generated on hills that are now almost entirely treeless (but used not to be).

In other parts of the world, river basin mismanagement has increased overland flow and flood risk greatly. Research on deforestation in Nepal shows a range of negative impacts linked with widespread loss of vegetation cover (Figure 18). Similar problems have arisen in Amazonia, where over 20% of the forest has been destroyed at an accelerating rate in the last 50 years by a combination of cattle ranching and large-scale agribusiness, such as soya bean production. As we shall see, river basin mismanagement on this scale has implications for the carbon cycle too (see p. 50, Figure 31).

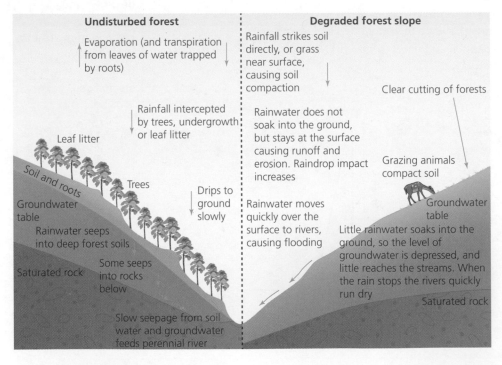

Figure 18 The impacts of deforestation on runoff generation

Summary

- There are three main causes of air uplift, condensation and cloud formation. These are the orographic, frontal and convection mechanisms.
- There are two recognised theories of precipitation formation. The Bergeron-Findeisen process explains why precipitation is able to fall rapidly after condensation begins in areas outside of the tropics. The collision process is more important for explaining sudden onset precipitation in the warmer tropics.
- Excess runoff (overland flow) generation occurs naturally in many parts of the world for climatic reasons. These include prolonged winter precipitation, monsoon rainfall and annual snowmelt.
- Increasingly, human factors are to blame for overland flow generation in many drainage basins. The two most important causes are urbanisation and forest removal.

Deficit in the water cycle

Meteorological causes of deficit in the water cycle

The balance between precipitation, evapotranspiration and runoff is known as the **water balance**. A **water deficit** is said to exist when precipitation is lower than the combined loss of water due to evapotranspiration and runoff. The water balance for a country or region provides a useful indication of how water availability may vary throughout the year. Drainage basin water balances are usually expressed using the following formula:

$$P = Q + ET + \Delta S$$

where P is precipitation, Q is discharge, ET is evapotranspiration and ΔS is positive or negative changes in water storage.

Figure 19 shows the monthly water balance for a small drainage basin in northern Canada. Important questions which you might attempt to answer in relation to this figure are:

- Do inputs of rainfall and snowfall equate with outputs of runoff and evaporation over the year as a whole?
- Why might water inputs greatly exceed outputs in January and February?
- Why do you think water outputs greatly exceed inputs in June (and to a lesser extent May)?

Drought conditions

At certain times and in some places, a water deficit may become serious enough for a condition of drought to be declared. There is no universally agreed specific definition of drought, but rather dozens of more technical measures that are used around the world. Three widely used definitions that additionally incorporate social and economic water deficit measures are:

- **Meteorological drought:** an extended period of low or absent rainfall relative to the average for a region.

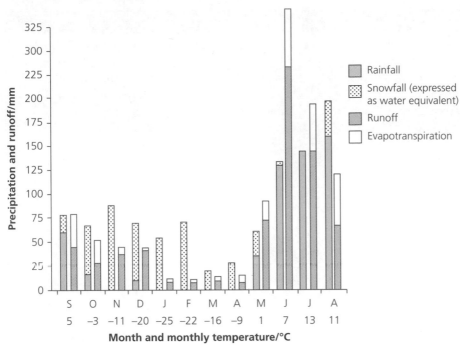

Figure 19 Annual water budget data for a small drainage basin in northern Canada

Self-study task 7

Study Figure 19. Describe the catchment inputs for (a) November and (b) June. To what extent can this drainage basin be viewed as a system in a state of equilibrium?

Fieldwork

A water cycle independent investigation can make use of qualitative data, such as the cartoon shown in Figure 20 or old newspaper reports of droughts. Interviews with older residents of a particular drainage basin may help uncover anecdotal evidence of major floods or excessive runoff events in the distant past.

- **Agricultural drought:** when there is insufficient moisture for average crop production.
- **Hydrological drought:** when available water reserves (in lakes, reservoirs and groundwater stores known as aquifers) fall below acceptable levels. This condition can arise even when there have been recent rains.

The 1976 UK drought

The protracted hydrological drought that affected the UK in 1976 had primarily meteorological causes.

- As a whole, the years 1975 and 1976 were exceptionally hot and dry. In both years, rainfall across the Midlands and southern England was around 50% of the average annual amount. In the Isle of Wight, only 20% of the normal rainfall fell between April and August 1976. The longest consecutive number of days with no rain was 45 (recorded at Milton Abbas, Dorset).
- In Cheltenham, the temperature exceeded 32°C for seven successive days. This is without parallel anywhere in the UK in recent times.
- Alongside quantitative data, old newspaper reports provide qualitative evidence for the severity of the water deficit. Contemporary accounts reveal that a Minister for Drought was appointed to send 'hosepipe patrols' in search of people breaking a national hosepipe ban. A public campaign promoted the practice of 'bathing with a friend', while newspaper cartoons of the time tried to see the funny side of a severe water deficit (Figure 20).

The explanation for the exceptional low rainfall was a northwards shift of the **jet stream**, which persisted for 18 months. The weather systems that generate frontal rain (see p. 28) follow the jet stream, and so much of the UK's normally expected rainfall was diverted northwards throughout much of 1975–76.

The **jet stream** is a narrow band of fast-moving winds high up in the atmosphere. It passes over a region of the north Atlantic where cold air from the Arctic meets warm air from the tropics, before reaching the UK. Rain-bearing weather fronts follow the course of the jet stream.

This was the driest 16-month period since 1727 for southern England. So unusual was the drought that meteorologists at the time calculated there was only a one-in-200 chance of it occurring again in any given year. However, global climate change (see p. 14) may mean the meteorological conditions that gave rise to the 1976 drought could recur more frequently in the future, perhaps once every 50 or 30 years (rather than every 200). The Intergovernmental Panel on Climate Change's projections suggest we will see increasing frequency and severity of drought in some parts of the world, including southern England (see p. 14).

Figure 20 A newspaper cartoon from 1976 provides qualitative evidence for a severe water deficit

Human causes of deficit in the water cycle

The human causes of a water deficit can include **aquifer** depletion and the over-extraction of surface water resources. Non-physical pressures that lead to these outcomes in some local contexts include population growth, agricultural demands and the use of water by industry (Table 5). Total freshwater withdrawals are predicted to increase by 2025 in all world regions.

An **aquifer** is a permeable rock layer which contains water that can be extracted for human use.

- In 1900, the total freshwater withdrawal figure was just under 600 cubic kilometres; by 1950 it had more than doubled.
- Total demand for water is projected to exceed 5,000 cubic kilometres by 2025.

Table 5 Human pressures on water supplies that may give rise to water cycle deficits

Agriculture, food and drink production	■ Crop cultivation, crop processing, food distribution (and even the final recycling phase for food and drink packaging) all require water ■ Agriculturally driven water stress is especially evident in the Indus river basin, which is home to the world's largest irrigation system. Irrigation accounts for over 90% of water withdrawn from available sources for use in many semi-arid regions of the developing world. Water stress in southeast Spain is also very high on account of fruit and vegetable production destined for European supermarkets ■ The Ogallala aquifer, which stretches from Texas to South Dakota, is being lowered at a rate of 90–150 cm per year. This will threaten one-third of irrigated agriculture in the USA within the next 40 to 180 years, with huge impacts on grain supplies and prices ■ In the drought-prone Indian state of Kerala, an aquifer lies close to the village of Plachimada. In 2000, Coca-Cola's subsidiary firm, Hindustan Coca-Cola Beverages, established a bottling plant near Plachimada. Six wells were dug, tapping into the precious groundwater store. Soon afterwards, water shortages began to be reported
Industry	■ After agriculture, industry is the second largest user of water ■ Growth in emerging economies since the 1980s has increased the amount of industrial activity worldwide. According to World Trade Organization statistics, the value of world trade in industrial products rose from US$2 trillion in 1980 to US$18 trillion in 2011, which gives some idea of the extra pressure placed on water supplies in recent decades
Household water consumption	■ Poverty reduction and the growth of emerging economies has increased water consumption globally: more people than ever before expect to have unlimited access to safe water for drinking, bathing, showering and domestic appliances (washing machines); sanitation and sewerage systems also place a heavy demand on water supplies ■ Worldwide, 2.6 billion people lack sanitation. However, as poverty alleviation increases further, more people are expected to make the transition to greater affluence (with the expectation of clean fresh water 24 hours a day, 7 days a week). This inevitably means more water consumption in many parts of the world where there are already water shortages, including north Africa and the Middle East

The shrinking Aral Sea

The inland Aral Sea in Kazakhstan and Uzbekistan, which has all but vanished, is a well-known example of the mismanagement of surface water resources. More than any other water body in the world, it has come to epitomise the devastating economic and ecological effects of excessive demands placed upon freshwater stores.

■ The Aral Sea was once the world's fourth largest inland sea, covering an area of 68,000 square kilometres.

■ It has steadily shrunk in size since the 1960s, following a succession of Soviet government schemes that diverted much of the river water which fed originally into the Aral Sea.

■ Large amounts of water were taken to irrigate fruit and cotton farming in what was once a biologically unproductive region. The irrigation demands of the cotton accelerated the shrinkage of the Aral Sea in the 1990s. Today, it covers only 10% of its original area.

In an ongoing effort to save the Aral Sea, Kazakhstan has secured World Bank loans in order to try to restore this important water resource. However, as Figure 21 shows, there is much work that needs to be done.

> **Self-study task 8**
>
> Describe the changes shown in Figure 21.

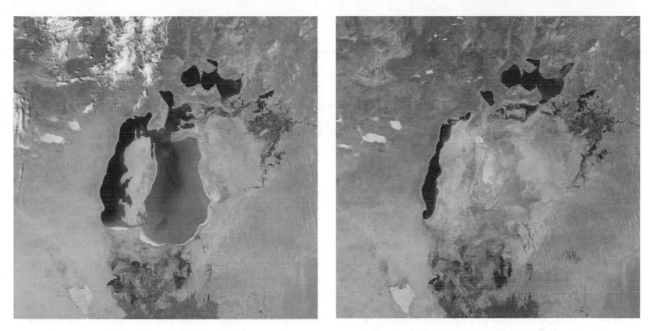

Figure 21 Contrasting satellite images showing how the Aral Sea reduced in size between 2000 and 2015

Natural and artificial recharge of aquifers

Aquifers are permeable and porous water-bearing rocks such as chalk and New Red Sandstone. The term is also applied to any water-saturated layer of gravel that has sufficient porosity and permeability to yield groundwater from wells and springs. Aquifers are naturally recharged over time by percolating rainwater. However, if too much water is taken from them too quickly then they will dry out before they have a chance to recharge. Many major cities such as Mexico City are located above groundwater aquifers. However, over-abstraction (over-pumping) sometimes takes place. When water is removed from the pore spaces in permeable rock, the rock loses strength. If too much weight has been added above by buildings, empty pores close up (imagine pressing down on a Swiss cheese full of holes filled with nothing but air). Ground beneath the city becomes compacted, the result being a permanent loss of water storage capacity.

Artesian aquifers develop where sedimentary rocks have formed a syncline (or basin-like) structure, with the aquifer confined between impermeable rock layers. The geology of the London Basin provides a good example of this and also of the issues associated with high rates of abstraction.

- The confined artesian aquifer below London consists of a saturated layer of chalk which is sandwiched between two layers of impermeable materials (London clay and Gault clay).
- Rainwater enters the chalk aquifer where it outcrops on the edge of the basin in the North Downs and Chilterns. Groundwater then flows by gravity through the chalk towards the centre of the basin.
- The level to which the water rises naturally is determined by the height of the water table in areas of recharge on the edges of the basin. If the groundwater is tapped by a well or borehole, water will flow to the surface under its own pressure.

Over-exploitation of London's aquifer in the nineteenth century (before modern piped water supplies were available) caused a drastic fall in the water table. In central London, it fell by nearly 100 m. In recent years, however, declining industrial demand for water and especially reduced rates of water abstraction using wells have allowed the water table to recover. By the early 1990s, the water table was rising at a rate of 3 m per year. As a result, water began to seep into the foundations of buildings and underground railway tunnels, which had been excavated in the past when the water table was lower. Measures have now had to be taken to prevent the natural recharge of the aquifer and to maintain the artificially low level that was established in the nineteenth century. The Thames Water company has been granted permission to abstract water from the aquifer to prevent any further rise of the water table.

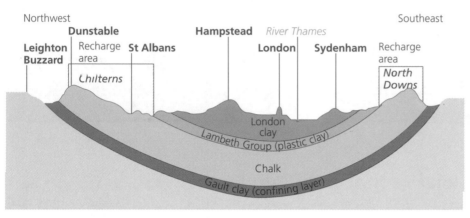

Figure 22 Geological section through the London Basin

Elsewhere in the world, several methods are used to recharge and restore aquifers to their natural state. If the ground surface is permeable, then water can be spread on the surface and allowed to percolate downwards. In the case of an artesian aquifer, water must be pumped into wells or deep pits (this method is used in parts of Israel).

Knowledge check 7

Explain how an aquifer develops. Why is human pressure on aquifers increasing in many parts of the world?

Summary

- A short-term water deficit can develop in a drainage basin for meteorological reasons, such as a low rainfall season. The water balance equation can be used to study periods of water deficit and surplus.
- A seasonal water deficit may last for an unusually long period of time because of naturally occurring variations in meteorological processes. The 1976 UK drought is an example of this. In the future, climate change could mean such events recur more frequently.
- Human pressures on aquifers and surface water resources are increasing because of rising affluence in emerging economies and population growth in some parts of the world.

- Industry and agriculture are responsible for aquifer and surface water depletion in some local contexts. The Aral Sea is a fraction of its former size because of over-use.
- Aquifers recharge naturally, provided that no permanent harm has been done to them (e.g. when over-extraction of water causes ground subsidence). Measures have been taken to prevent recharge of London's aquifer in order to stop the underground railway system from flooding. Elsewhere, artificial recharge is achieved by various methods.

■ The global carbon cycle

Carbon inputs, outputs, stores and flows

Carbon is ubiquitous (meaning it is found everywhere). Thanks to its ability to bond with many other elements, it has become the 'building block' for life on Earth.

- Carbon-based molecules are integral to all living creatures.
- Carbon dioxide (CO_2) and methane (CH_4), another atmospheric gas containing carbon, are parts of the air we breathe.
- Carbon exists in a dissolved form in water, and is present in limestone, fossil fuels, ocean sediments and soils.

Carbon has become important politically too, because of the role CO_2 plays in anthropogenic climate change.

Like water, carbon is continually cycled at the global scale. When local-scale studies of the **carbon cycle** are undertaken (for instance, a local woodland investigation), the local flows and stores that are explored can be viewed as a subsystem which also belongs to the bigger picture of global cycling and storage.

In parallel with water cycle studies, we can distinguish between important carbon system elements:

- **Stores** (or stocks) are the amounts of carbon held in a particular part of the global system. For example, around 2,300 **gigatonnes** of carbon are stored in soil and **peat** worldwide. The **residence time** of carbon is the average length of time it remains in any carbon store.
- **Flows** are movements or transfers of carbon between the stores. For example, volcanic activity adds 0.1 gigatonnes of carbon to the atmosphere every year. The rate of flow (measured as units of mass per unit time) is sometimes called a flux.
- Flows into or out of particular stores are called **inputs** and **outputs**. For example, since 1750 new inputs of carbon dioxide and methane have been added to the atmospheric carbon store by human activities, including fossil-fuel burning, deforestation, land use changes and cement manufacturing. Before the Industrial Revolution, the concentration of carbon dioxide in the atmosphere store was approximately 280 parts per million (ppm). Over the last 250 years, this concentration has increased at an accelerating rate and now exceeds 400 ppm.
- **Processes** are the physical mechanisms that drive the inputs, outputs and flows of the cycle. For example, **photosynthesis** is a key process that drives the flow of carbon from the atmosphere to the vegetation store. Chemical processes and water flows combine to drive the flow of carbon from the lithosphere (land) to the ocean store (via runoff and river flow).
- The concept of **mass balance** can be applied to the carbon cycle: this means that at a global scale, the total amount of carbon is conserved over time (although long-term changes may occur in where it is stored as a result of climate change and sedimentary processes).

The **carbon cycle** is the biogeochemical cycle by which carbon moves from one part of the global system to another. At the global scale, it is a closed system made up of linked inputs, outputs, flows and stores. At the local scale, it is an open system.

One **gigatonne** amounts to 1 billion tonnes (or 1 trillion kilograms). A gigatonne of carbon dioxide equivalent (GtC) is the unit used by the United Nations climate change panel, the Intergovernmental Panel on Climate Change (IPCC), to measure the amount of carbon in various stores.

Peat is partly decomposed organic matter that has accumulated in waterlogged conditions.

Photosynthesis is the process whereby plants use sunlight to combine carbon dioxide (CO_2) from the air with water and minerals to make complex molecules.

Figure 23 shows the global carbon cycle in diagrammatic ('black box') form and Table 6 provides a summary of the main global carbon stores. Carbon flows through these stores are driven by a range of natural processes which are explored in the following pages.

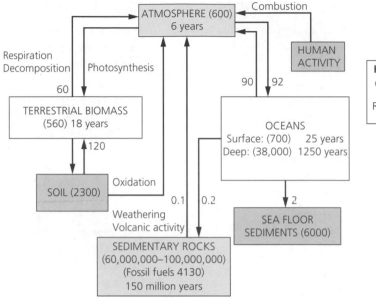
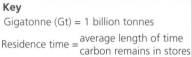

Figure 23 The global carbon cycle: flows, driving processes, stores and residence times

Table 6 Global carbon stores (figures are naturally occurring volumes and do not take into account human activity)

Store	Carbon in store (gigatonnes)
Sedimentary (carbonate) rocks and deep ocean sediments	100,000,000
Ocean water (biomass and dissolved CO_2)	38,700
Sea floor sediments	6,000
Fossil fuels (coal, oil, gas)	4,130
Soils and peat	2,300
Atmosphere (gaseous CO_2)	600
The biosphere (forests and grasslands)	560

Carbon pathways and processes

Fast carbon cycle flows between the land and the atmosphere

At a more short-term and local scale, several important processes are involved in carbon sequestration and release. These are shown in Table 7 and are sometimes referred to as 'fast carbon cycle' processes.

Carbon sequestration is the natural capture and storage of carbon dioxide (CO_2) from the atmosphere by physical or biological processes such as photosynthesis.

Exam tip

Make sure you can use data to support an overview of the various global carbon stores. For instance, the atmosphere naturally contains around 600 billion tonnes of carbon, or 600 GtC (as a result of human activity, however, the figure has actually increased to more than 800 GtC).

Fieldwork

If you carry out a carbon-based local-scale independent investigation, you will almost certainly focus on fast carbon cycle flows and storage.

Net primary productivity (NPP) is the accumulation in dry weight of green plant material per unit area per unit time.

Table 7 Relatively fast carbon cycle flows between the land and the atmosphere

Process	How it works
Photosynthesis	■ This is the process by which carbohydrate molecules are produced from carbon dioxide and water using energy from light. As they photosynthesise, plants 'fix' gaseous carbon dioxide from the atmosphere (Figure 24) into solid form in their living tissues as part of the process. Oxygen is a byproduct that is released into the atmosphere ■ The photosynthesis process operates unevenly throughout the day and year; plant communities are far more productive in some climatic regions of the world than others. Carbon flows between the atmosphere and the biosphere therefore vary greatly both temporally and spatially ■ The net primary productivity (NPP) of an ecosystem is a measure of the rate at which new organic matter is produced by photosynthesis per unit area per unit time. The three global environments with the highest rates of NPP are, in order: shallow warm-water estuaries, marshes and tropical rainforests
Respiration	■ Carbon dioxide is released back into the atmosphere by living organisms through the process of respiration. Plants create the energy they need for respiration by breaking down stored glucose (sugars). Carbon dioxide is given off as a byproduct. Essentially, respiration reverses the photosynthesis process ■ Like photosynthesis, rates of respiration vary between different world biomes and throughout the day and year in particular places ■ Together, photosynthesis and respiration are vital processes driving the fast carbon cycle. Photosynthesis removes carbon from the atmosphere, while respiration replaces it. However, these processes are not in complete balance. Some of the organic matter that is produced by photosynthesis becomes buried in sedimentary rocks, or is stored naturally deep underground as fossil fuel. Over geological time, more carbon dioxide has been removed by photosynthesis than has been released through respiration
Decomposition	■ CO_2 from plants and animals is also returned to the atmosphere through processes of dead tissue decomposition. When living organisms die, their cells begin to break down as a result of physical (wind and water), chemical (leaching and oxidation) and biological (feeding and digestion) mechanisms. The last of these is carried out by microorganisms such as bacteria and fungi: these specialist organisms break down the cells and tissues in dead organisms. Carbon dioxide is released during this process ■ Rates of decomposition vary greatly from place to place and time to time. In humid locations with warm temperatures all year round, such as tropical rainforests, dead plants and animals can decompose beyond recognition within days. In contrast, decomposition occurs so slowly in the Arctic tundra that recognisable plant and animal remains may lie on the surface of the ground for many months, or even years
Fossil fuel combustion	■ Fossil fuel (hydrocarbon) combustion takes place rapidly in the presence of oxygen and releases carbon dioxide. Industrial societies burn coal, gas and oil to provide the energy they need for electricity and transport, thereby transferring large amounts of stored carbon into the atmosphere, some of which is subsequently absorbed by the oceans. Around 85% of global energy consumption is derived from fossil fuel use despite recent advances in renewable energy such as wind and solar power ■ Traditional societies burn biomass on demand to provide warmth and a heat source for cooking

Carbon cycle flows from the atmosphere to the ocean

The oceans take up carbon dioxide by carbon cycle 'pump' mechanisms (Figure 25).

■ The physical (inorganic) pump involves the movement of carbon dioxide from the atmosphere to the ocean by a process called diffusion. CO_2 dissolved in the surface of the ocean can be transferred to the deep ocean in areas where cold dense surface waters sink. This downwelling carries carbon molecules to great depths where they may remain for centuries. The level of CO_2 diffusion also determines the acidity of the oceans (see p. 61).

Biomass is the biological material derived from living or recently living organisms; or the weight of dry organic matter per unit area in an ecosystem.

Carbon cycle pump processes operate in oceans to circulate and store carbon.

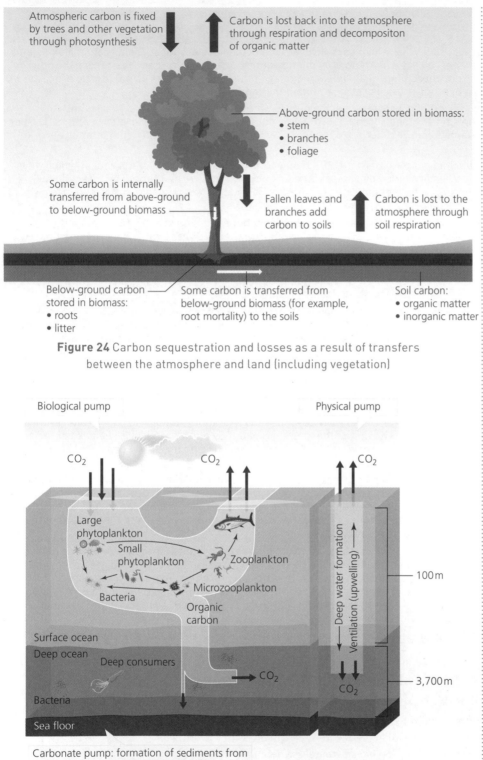

Atmospheric carbon is fixed by trees and other vegetation through photosynthesis

Carbon is lost back into the atmosphere through respiration and decompositon of organic matter

Above-ground carbon stored in biomass:
• stem
• branches
• foliage

Some carbon is internally transferred from above-ground to below-ground biomass

Fallen leaves and branches add carbon to soils

Carbon is lost to the atmosphere through soil respiration

Below-ground carbon stored in biomass:
• roots
• litter

Some carbon is transferred from below-ground biomass (for example, root mortality) to the soils

Soil carbon:
• organic matter
• inorganic matter

Figure 24 Carbon sequestration and losses as a result of transfers between the atmosphere and land (including vegetation)

Biological pump

Physical pump

CO_2

CO_2

CO_2

Large phytoplankton

Small phytoplankton

Zooplankton

Bacteria

Microzooplankton

Organic carbon

Surface ocean

Deep ocean

Deep consumers

CO_2

Bacteria

Sea floor

Deep water formation

Ventilation (upwelling)

CO_2

100 m

3,700 m

Carbonate pump: formation of sediments from dead organisms. Sedimentation sequesters

Figure 25 Oceanic carbon pumps

Self-study task 9

Explain the different ways in which carbon is transferred from the atmosphere to the ocean over varying timescales.

- The biological (organic) pump is driven by ocean **phytoplankton** absorbing carbon dioxide through photosynthesis. These organisms form the bottom of the marine food web, and they live in the ocean's surface layer (euphotic zone) where sunlight penetrates. Phytoplankton are consumed by other marine organisms and carbon is subsequently transferred along food chains by fish and larger sea animals as they consume one another.
- Organic carbon may eventually be transferred to the deep ocean when dead organisms sink towards the ocean floor (this is called the carbonate pump).

In summary, both very fast (photosynthesis) and slower (downwelling) processes are involved in the flow of carbon into and out of ocean storage.

> **Phytoplankton** are tiny, sometimes microscopic, plant organisms that float and drift in the oceans, capturing the sun's energy through photosynthesis.

Slow carbon cycle flows between the land and the ocean

This is sometimes called the 'slow carbon cycle'. The cycling of carbon between bedrock stores on the land and the oceans occurs through processes of weathering, erosion and deposition over very long timescales (millions of years), and at a continental scale (Figure 26).

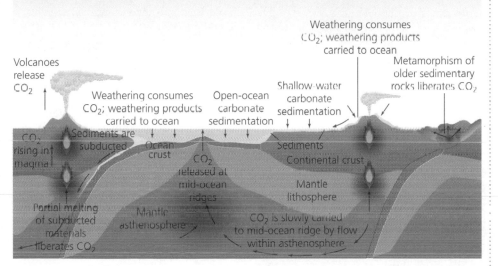

Figure 26 The slow geological (land–ocean) carbon cycle (source: D. Bice, The long-term carbon cycle)

First, **weathering** of rocks takes place on the land. Most chemical weathering processes involve rainwater containing dissolved carbon dioxide which has produced weak carbonic acid (H_2CO_3). **Carbonation** is the principal chemical weathering process affecting carbonate rocks such as chalk and carboniferous limestone. These rocks are composed mainly of calcium carbonate which reacts with carbonic acid to produce calcium bicarbonate, which is soluble.

Globally, some 0.3 billion tonnes of carbon are transferred from rocks to the atmosphere and oceans each year by chemical weathering. The removal of carbon in solution from the land is an important way in which the carbon cycle and water cycles are interrelated. Calcium bicarbonate can be transferred by overland flow, throughflow or groundwater flow into rivers where it becomes part of the **solute load** (dissolved material carried in solution, an important river transport mechanism).

> **Weathering** is the in situ breakdown of rocks at or near the Earth's surface by physical, chemical and biological processes.

> **Carbonation** is a type of chemical weathering of rocks by rainwater which, in combination with dissolved carbon dioxide, forms a weak carbonic acid. This changes any rock minerals that contain lime or other basic oxides into soluble bicarbonates.

- Over time, large amounts of carbon have been removed in solution from limestone areas of the UK, such as the cliffs at Llandudno.
- Carbonation weathering takes place when rainwater collects in pools on the surface of exposed rock, for example the limestone pavement at Malham, Yorkshire.
- In chalk regions such as England's South Downs, the slow movement of groundwater dissolves the rock it is transmitted through, and eventually transfers calcium bicarbonate in solution into river systems and ultimately the ocean.

Once in the ocean, carbonate is used by marine organisms to create shells. When they die, these organisms' carbonate shells are deposited as carbonate-rich sediment on the ocean floor where they are eventually **lithified** (turned into rock). This part of the carbon cycle can lock up carbon for millions of years. It is estimated that the oceanic sedimentary layer may store up to 100 million GtC.

- Some carbon is eventually returned to the atmosphere by volcanism, as CO_2 is released from melted rocks when subduction occurs at plate boundaries.
- Figure 26 shows how huge volumes of stored carbon are constantly on the move (albeit at extremely slow rates of movement) through the geological process of tectonic plate movement.

One of Earth's largest carbon stores is the Himalayas. These mountains are composed of oceanic sediments rich in calcium carbonate. Folded upwards long ago by mountain-building tectonic processes, this carbon is now being actively weathered, eroded and transported back into the oceans. Water cycle processes including monsoon rainfall and runoff play an important role in this.

Exam tip

Make sure you have learned and know how to use specialist carbon process terminology, including sequestration, photosynthesis, respiration, glucose, phytoplankton and carbonation.

Knowledge check 8

What is the cause of subduction at plate boundaries, and how can it lead to the melting of carbonate rocks?

Exam tip

Use examples of specific landscapes, such as Llandudno or the Himalayas, from which carbon is removed in solute form by the combined processes of weathering and water transport.

Self-study task 10

Construct a revision table showing how different carbon pathways and processes operate at varying spatial and temporal scales.

Summary

- The global carbon cycle consists of numerous stores, flows, inputs and outputs. Enormous amounts of carbon are stored in sedimentary rocks and deep ocean sediments.
- Photosynthesis, respiration, decomposition and fossil fuel combustion are important processes controlling carbon flows between the atmosphere and the land. These processes can be studied at the local level over very short time periods.
- Carbon cycle flows from the atmosphere to the ocean take place via two carbon cycle pump mechanisms, known as the physical pump and the biological pump. Marine organisms — especially phytoplankton — drive the biological pump.
- A slow carbon cycle operates between the land and the ocean over thousands or even millions of years. Numerous processes are involved including carbonation weathering, river transport of carbonates in solution, the lithification of carbon-rich sediments and plate tectonic movement and melting.

■ Carbon stores in different biomes

Ecosystem carbon storage

When human activity is taken into account, the total amount of carbon stored in the terrestrial biosphere is estimated to be approximately 3,000 GtC. This storage is spread unevenly among the different terrestrial biomes, as Figure 27 shows. Forests are significant carbon stores; they make up more than half of all terrestrial ecosystem storage. Carbon is stored primarily in the biomass of the trees but a thick litter layer on the forest floor can also store significant amounts.

- Boreal (coniferous) forest stores more carbon than tropical rainforest globally because it is distributed across a greater area, including much of northern Russia and North America.
- The biomass of tropical rainforest (the weight of organic matter per unit area) is greater than that of boreal forest but it does not cover as great a land area.

In order to grow and create new organic matter, plants require a variety of nutrients. Some are provided by the photosynthesis process while others are extracted from the soil by plant roots (using a process called cation exchange, for example). Carbon is an essential plant macronutrient and makes up approximately 44% of the dry weight of plant biomass. This is because it is a major component of all organic molecules. By comparison, nutrients such as potassium and calcium may amount to less than 1% of dry weight.

Figure 28 shows the carbon flows and stores in a biome. The proportional sizes of the plant, animal, litter and soil stores vary greatly between biomes and may also change within a particular ecosystem according to season. Note that the carbon shown in the soil in Figure 28 is that which is contained in organic matter (and not inorganic carbon derived from the underlying bedrock and sediments).

- **Green plants:** at the global scale, nearly 20% of the carbon in the Earth's biosphere is stored in plants. Although the exposed part of the plant is the most visible, the below-ground biomass (the root system) must also be considered. In grasslands, the majority of the biomass consists of root systems.
- **Animals:** these play a small role in the storage of carbon. This is because the biomass of the animals in a biome is much smaller than that of plants and because of the inefficient energy transfers that take place between different trophic levels.
- **Litter:** this is defined as fresh, undecomposed, and easily recognisable plant debris. This can include leaves, pine needles and twigs. Leaf tissues account for about 70% of litter in forests.
- **Soil:** humus is a black substance that remains in the soil after most of the organic litter has decomposed. It gets dispersed throughout the soil by earthworms and other organisms.

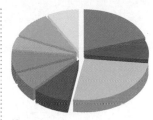

Key
- ■ Tropical forest 20%
- ■ Temperate forest 7%
- ■ Boreal forest 26%
- ■ Agriculture 9%
- ■ Wetlands 7%
- ■ Tundra 8%
- ■ Desert 5%
- ■ Temperate grassland 10%
- ■ Tropical savanna 8%

Figure 27 The carbon storage contribution of different world biomes

Knowledge check 9

How are energy and matter transferred between trophic levels in an ecosystem? What factors may influence how much carbon is transferred between trophic levels?

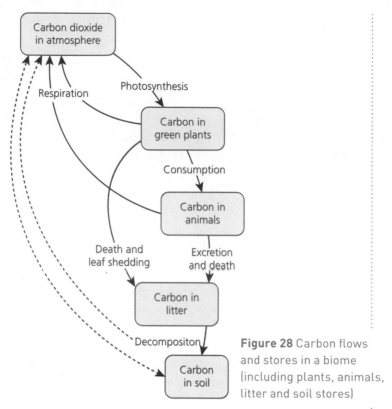

Figure 28 Carbon flows and stores in a biome (including plants, animals, litter and soil stores)

Tropical rainforest carbon storage

Figure 29 shows the global distribution of biomes. Tropical rainforests are found on and around the equator in Asia, Africa and South America, between 10°N and 10°S latitude, at elevations below 1,000 m. There are three major formations:

- Amazonian (Amazonia into Central America)
- African (Zaire Basin and West Africa; also Madagascar)
- Indo-Malaysian (west coast of India, southeast Asia, New Guinea and Queensland, Australia)

Table 8 outlines factors affecting plant growth and carbon storage in tropical rainforest regions.

As a result of the conditions shown in Table 8, rainforest net primary productivity (NPP) averages around 2,500 grams per square metre per year; the biomass can be as much as 700 tonnes per hectare.

- Competition for light and water has given rise to five layers of vegetation, including a field layer, shrub layer, under-canopy and continuous canopy (around 30 metres high) from which emergent trees project even higher into the sky.
- Plants are evergreen and the trees have large leaves that maximise their rate of transpiration and growth.
- Carbon storage in animals is relatively high in this ecosystem because of the large number of habitats provided for plants, insects, frogs, birds and mammals. More than two-thirds of the world's plant species are found in the world's rainforests. One study in the Napo region of Peru, for example, found 283 species of trees in an area of forest no larger than a football pitch.

> **Self-study task 11**
>
> Compare the global distribution of tropical rainforest and temperate grassland.

Table 8 Physical factors affecting plant growth and carbon storage in the tropical rainforest biome

Factor	Influence on plant growth and carbon storage
Light	The sun's rays are concentrated at this latitude, with little seasonal variation. This results in all-year growth and carbon sequestration
Temperature	High average annual temperatures between 25°C and 30°C with only small seasonal variations
Precipitation	The sun's heating causes moist air to rise which leads to heavy rainfall most days, and a high average annual total of 2,000–3,000 mm with no dry season (although some locations, such as Manaus in Figure 29, experience a few months of the year when rainfall is lower). High and constant rainfall combined with warm temperatures provides optimum conditions for terrestrial plant growth and biomass carbon storage

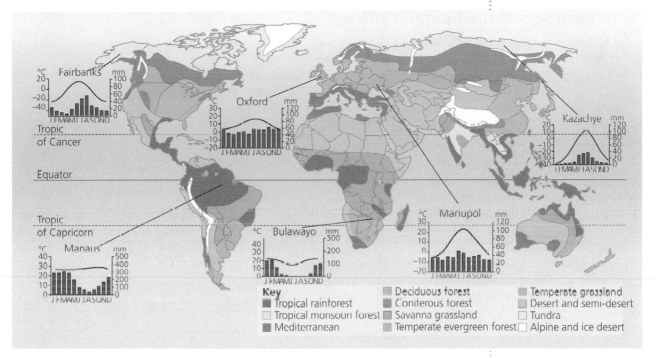

Figure 29 The global biome map

Rainforest carbon storage and flows

In total, 550 GtC is stored in tropical rainforest biomass and soil.

- Large forest trees typically store 180 tonnes of carbon per hectare above ground and a further 40 tonnes of carbon per hectare in their roots. The soil carbon store averages around 100 tonnes per hectare.
- Compared with other forests, exchanges of carbon between atmosphere, biosphere and soil are rapid. Warm, humid conditions ensure rapid decomposition of dead organic matter and the quick release of CO_2. As a result, the litter store is proportionately small in this biome. Figure 30 shows the proportional size of the carbon stores in tropical rainforest and temperate grassland biomes.
- Heavy rainfall means that soils are leached and only retain limited amounts (proportionately) of organic carbon in the form of humus.

Self-study task 12

Compare the levels of carbon storage in other biomes using the map at www.carbon-biodiversity.net/Issues/CarbonStorage

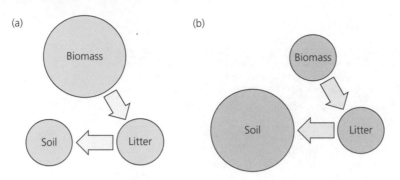

Figure 30 The proportional size of carbon-containing nutrient stores in (a) tropical rainforest and (b) temperate grassland biomes

Temperate grassland carbon storage

Mid-latitude temperate grasslands are found (1) on the periphery of mid-latitude deserts, typically between 30° and 50° north of the equator (only small pockets lie south of the equator) and (2) the leeward side of mountain systems which act as a barrier to westerly flowing moist air, thereby giving rise to a rain shadow (see p. 28). Their global distribution — shown in Figure 29 — comprises several extensive non-coastal regions:

- North American **prairies** (Central Lowlands and High Plains of the USA and Canada)
- the Eurasian **steppes** (from Ukraine eastward through Russia and Mongolia)
- South Africa's **veld** landscape
- the South American **pampas** (Argentina and Uruguay)

Table 9 outlines factors affecting plant growth and carbon storage for this biome.

The lack of rainfall is a major limiting factor preventing the growth of thick forest cover. Instead, short perennial (long-lived) grasses dominate the landscape. Grasses, with their growth buds at or just below the surface, are well-adapted to drought, fire

Table 9 Physical factors affecting plant growth and carbon storage in the temperate grassland biome

Factor	Influence on plant growth and carbon storage
Light	The sun's rays are concentrated during summer at these latitudes but are much weaker during winter months when there may be no more than 6 hours of daylight. This results in marked seasonal variations in plant growth and biomass carbon storage
Temperature	The mean monthly temperature in Mariupol (Ukraine) varies between 22°C in summer and −5°C in winter. Temperate grassland at high altitude or closer to the coniferous forest boundary may experience an even more extreme temperature range
Precipitation	Low average annual rainfall of around 500 mm or below; in much of the biome (including the grassland near Mariupol) this is spread relatively evenly throughout the year. However, there is a substantial build up of winter snow in temperate grasslands found at higher northern latitudes

Knowledge check 10

Why does the number of daylight hours vary so greatly between seasons at high latitudes?

and cold. Their narrow, upright stems reduce heat-gain in the hot summers; their intricate roots trap moisture and nutrients below ground. Two basic types are:

1 turf grasses, with rhizomes (or underground stems) from which new plants grow. These are associated with more humid parts of the biome

2 bunch grasses, without rhizomes, that reproduce by seed. These are associated with the drier parts of the biome

Animal biodiversity and biomass is relatively low in this structurally simple climate and vegetation zone. For example, usually no more than two or three species of large grazing mammals, such as bison, occupy a typical temperate grassland. Below ground is a different story: the majority of temperate soils support significant earthworm populations, and 1 square metre may contain up to 500 earthworms. As they consume organic matter and excrete waste, earthworms play a key role in carbon cycle movements for temperate grasslands. In just 10 years, the entire top 15 cm of soil may be passed through the guts of earthworms.

Grassland carbon storage and flows

In total, 185 GtC is stored in temperate grassland biomass and soil.

- Temperate grassland typically stores 2–10 tonnes of carbon per hectare above ground. Around double this amount is stored additionally below ground as roots (in most temperate grassland ecosystems, two-thirds of the root biomass is found in the top 30 cm of the soil).

- The soil carbon store (consisting of humus, as opposed to living roots) averages around 100–200 tonnes of carbon per hectare.

- Warm, humid conditions in autumn ensure rapid decomposition of dead grass matter and the quick release of CO_2. As a result, the litter store is small in this biome.

- Exchanges of carbon between atmosphere, biosphere and soil vary greatly according to season. In winter, the grasses die back to their roots and photosynthesis ceases. Plants can continue to respire through their roots, however, especially once the soil begins to warm in spring.

Changes in biome carbon stores due to human activity

Changes in the size of biome carbon stores may result from a range of human activities. The land use changes of deforestation, afforestation and agricultural activity all affect carbon sequestration.

Deforestation

Forest landscapes are important resources for human populations. Rainforest timber harvesting — for example, mahogany and teak — has been occurring for centuries to meet demand for tropical hardwood furniture and flooring. This trade continues today, despite attempts to restrict imports by countries such as the UK. In recent decades, large-scale land use changes carried out by agribusinesses have accelerated rainforest removal in Amazonia, central Africa and Indonesia.

- Huge areas of Amazonian forest have been cleared for commercial agriculture. Deforestation in Amazonia averaged around 17,500 square kilometres per year between 1970 and 2013. Major crops such as soya beans are grown on old rainforest soils in Brazil. In neighbouring Costa Rica, around one-third of all cleared rainforest land is used for cattle ranching. Rising demand for food comes from an increasing large and affluent global population.

■ Oil palm cultivation takes place at the expense of tropical rainforest cover. In recent years, forest fires in Sumatra and Kalimantan (Indonesia) have caused a toxic haze that spread across a large part of southeast Asia. Clearance of tropical forests for settlement and agriculture has led to significant increases in carbon emissions.

Deforestation removes the carbon biomass store. Croplands and pasture contain just a fraction of carbon stored in trees. For example, the biomass of soya crop cover is 2.7 tonnes per hectare (compared with 180 tonnes per hectare for virgin rainforest). At the same time, deforestation reduces inputs of organic matter to the soils. Moreover, increased overland flow can result in soil erosion and the permanent loss of carbon storage capacity (this demonstrates another way in which the water and carbon cycles are interrelated). Figure 31 shows some of the impacts of deforestation on the carbon cycle.

It is important to note, however, that the above-ground biomass of surviving rainforest is believed to have *increased* in size slightly in recent years. This is thought to be due to sequestering of increased CO_2 concentrations in the atmosphere and is an example of negative feedback in the carbon cycle system (see p. 64).

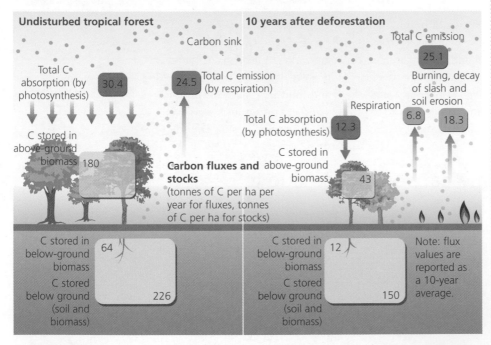

Figure 31 The impact of deforestation on the carbon cycle over a ten-year period

negative feedback in the carbon cycle system (see p. 64).

Self-study task 13

Using data from Figure 31, compare the relative size of different carbon flows and stores before and after deforestation has taken place.

Afforestation

Afforestation involves planting trees in deforested areas or places that have never been forested. New trees act as carbon sinks and can therefore help with **climate change mitigation** (see p. 45). Afforestation also helps to reduce flood risk, as we have seen (see p. 32).

- The UN's Reducing Emissions from Deforestation and forest Degradation (REDD) scheme provides incentives for developing countries to conserve their rainforests by placing a monetary value on forest conservation.
- The UK Forestry Commission was established in 1919, following the end of the First World War, to increase timber supplies through a policy of land use changes. Marginal areas of grassland, heather and moorland were used to grow coniferous forest, for example in the Brecon Beacons (Wales) and the Isle of Arran (Scotland). In recent years, the Forestry Commission has used carbon sequestration as a justification for its work, arguing that: 'Carbon sequestration is a pure international public good. It has equal benefit to everyone. One tonne of carbon locked up in Scotland has the same value as one tonne locked up in Wales, as a contribution to reducing global warming.'
- New **monoculture** of commercial trees, such as coniferous plantations in the UK, can increase carbon storage if it replaces grassland. However, it may store less carbon than natural forest biome communities do. This is because monoculture forest lacks biodiversity and provides few habitats for other plant and animal species to occupy.
- Individual citizens can play an active role in afforestation through the practice of **carbon offsetting**, which is another widely used mitigation strategy that aims to marry business principles with environmental goals. Our everyday actions, such as driving, flying and heating buildings, consume energy and produce carbon emissions. Carbon offsetting is a way of compensating for your emissions by funding an equivalent carbon dioxide saving elsewhere. The suggestion that we 'plant a tree' after taking an aeroplane flight is a well-known example of the offsetting principle.

Agricultural activity

Agriculture affects both biomass carbon storage of land and the amount of soil organic carbon (SOC) that is present. SOC is the carbon associated with soil organic matter (made up of humus, i.e. decomposed plant and animal materials).

- Globally, clearing natural vegetation for agriculture brings a large reduction in SOC levels; further declines may occur because of poor management practices. In many farmed areas, soil carbon levels have fallen by 50% compared with pre-agricultural periods.
- Historically, excessive cultivation using inappropriate methods has resulted in soils being 'overworked'. The consequent loss of SOC has led to many land degradation and soil erosion problems.
- Soil erosion that occurred in the USA during the 1920s and 1930s (the 'Dust Bowl' episode) is a well-known example of this. In the Great Plains, wind erosion stripped the top soil from 65 million hectares of over-cultivated land, leading to an enormous loss of soil carbon storage capacity.

Positive changes in carbon storage can also result from agricultural activity, however (Table 10). These activities include the addition of manure, plant debris, composts and biosolids from sewage to agricultural soils. All are high in organic carbon and represent additional carbon inputs to the soil subsystem.

Climate change mitigation consists of any action intended to reduce GHG emissions, such as using less fossil fuel-derived energy, thereby helping to slow down and ultimately stop climate change. Mitigation can be practised by stakeholders at different scales, from a citizen switching off a light, to a government setting strict national targets for reduced carbon emissions.

The planting of a single species is called **monoculture**.

Table 10 Farm management practices that may increase soil carbon storage

Management category	Examples of good practices
Crop management	Irrigation Improved crop rotation
Pasture management	Fertiliser management Grazing management Earthworm introduction Improved grass species Introduction of legumes
Organic amendments	Animal manure Recycled plant remains

Summary

- Carbon is stored in the world's biomes in varying amounts. Climatic controls including temperature, precipitation and light are responsible for the relatively large biomass carbon store in tropical rainforests and the large soil carbon store in temperate grasslands.
- Land use changes affect the size of natural carbon stores. Deforestation has been occurring for thousands of years and has reduced carbon storage at both national and global scales.
- Afforestation can help with climate change mitigation by increasing carbon sequestration and storage. The UK first adopted widespread afforestation policies in 1919 and individuals can play a role too, for example by funding tree planting after taking an aeroplane flight.
- Agricultural activity has a wide range of effects on carbon storage and flows. The removal of natural forest cover and soil erosion following over-cultivation reduces carbon storage. However, a wide range of good management practices can help restore carbon to the soil store.

■ Changing carbon stores in peatlands over time

Peat formation and carbon storage accumulation

Peat is a thick layer of black or dark brown sticky and wet soil material. It is a substance that hill walkers, gardeners and farmers will be familiar with. However, unless you have spent much time in the countryside, you may not have encountered peat personally.

Large areas of **peatlands** exist in upland areas of the UK and some lowland regions, such as the poorly drained fenlands of East Anglia (Figure 32).

- The black or dark brown colour of peat derives from very high levels of only partially decomposed vegetable matter (derived from sphagnum mosses, rushes, sedges and bracken). The plant remains are slowly compressed as more material is added each year. Over time, layer upon layer of dead matter accumulates until, in the UK, depths of 2–4 m can be achieved (Figure 33). Decomposition is prevented by a waterlogged environment which creates oxygen-deficient **anaerobic** conditions.

Peatlands are landscapes where layers of peat have accumulated on the land surface. The organic content of the surface horizon exceeds 80% and the peat depth is at least 40 cm. Globally, peatlands can cover many tens or hundreds of square kilometres with thicknesses of 10 m or more.

Anaerobic conditions arise where there is no air and little oxygen, such as waterlogging in soil.

- This limits microbial decomposition of organic matter, especially in upland sites where temperatures are cooler and vegetation is in any case more acidic (which deters soil organisms).

- In many sites, peat has been accumulating for much of the **Holocene**, and most peatlands had begun to form by around 7,000 years ago. Occasionally, recognisable animal remains from thousands of years ago are found preserved in peat. In 1984, a 2,000-year-old well-preserved human body was found buried in Wilmslow, Cheshire (the media named him 'Pete Marsh').

The **Holocene** is the current geological era, which began around 10,000 years ago when the Pleistocene ended.

- Deep peaty soils
- Shallow peaty soils
- Soils with peaty pockets

Figure 32 The distribution of peat in England and Wales (prior to human impacts)

Figure 33 Where peat has been cut for fuel, its depth can be measured easily

Anaerobic conditions are the product of specific **hydro-topography** (a combination of drainage and relief factors). There are many names for poorly drained anaerobic sites, including marsh, swamp and bog.

■ **Fen peatlands** form where groundwater meets the surface — at springs, hollows or at the edge of open water.

■ **Blanket peatlands** occur on hill tops which receive over 1,500 mm of rain a year, are fed entirely by rainfall and snowmelt rather than groundwater, and are typically found in the wetter north and west of the UK. About 1.5 million hectares of blanket bog are found in the UK, the majority in Scotland.

■ **Raised bogs** occur in valley bottoms where soils are saturated because of the frequent arrival of throughflow and overland flow from the slopes above (see Figure 7, p. 19). Over time, the accumulation of peat naturally forms a dome-shaped raised area of land.

Carbon storage in peat

Some carbon is lost from peatlands as a result of water cycle movements (peatland river water can have a noticeably dark colour; in some peatland areas, even tap water may have a brownish tint). However, there is a net accumulation of carbon over time in undisturbed peatland, making it an important carbon store (the organic matter in peat is 50% carbon). This is because the rate at which atmospheric carbon is fixed in new peatland plants by photosynthesis greatly exceeds the rate of loss of carbon from the same local subsystem through the decomposition of plant litter. Instead, the plant litter accumulates on the ground surface or in the upper soil horizons (in the form of dead plant roots).

■ Over half of all UK soil carbon is stored in peatlands, because of its deep organic layer, and up to 90% of this is in Scotland. The carbon stored in UK peatlands is equivalent to 3 years of total UK carbon emissions.

■ Despite covering only 5% of the land area of Wales, deep peatland soils represent 30% of the country's total soil carbon storage.

■ Globally, peatlands store approximately double the amount of carbon that is stored in all the world's forests, an estimated 550 billion tonnes.

Fieldwork

Carbon stocks in peatlands could be the focus for an independent investigation. Peatland depth is easily measured by finding sites where peat cuttings are visible and measuring the depth. Where the age of local peatlands is known from secondary data, estimates can be made of its growth as a carbon store over time.

The global distribution of peat

In total, 3% of the Earth's land surface — 4 million km² — is covered by peat. Peat occurs on all continents, from tropical to Arctic zones and from sea level to high altitudes. More than 250 GtC is stored in peat globally.

■ 20% of global soil carbon is stored in extensive peatlands in Siberia and Canada.

■ Peat soils in some parts of the humid tropics are composed of the remains of rainforest trees. The tropical peatlands of Indonesia are a product of low relief, impermeable rock and high rainfall; they store 60 billion tonnes of carbon. Raised deposits of woody peat can reach depths of 15 m and store up to 3,000 tonnes of carbon per hectare.

Self-study task 14

Find out more about peat resources in Wales: www.forestry.gov.uk/fr/INFD-8Z7BSH or England: http://publications.naturalengland.org.uk/publication/30021

Exam tip

Peat formation is a relatively complex process and you will need to take care with terminology (such as sphagnum mosses and anaerobic conditions).

Peat extraction and drainage

The drainage, burning, cultivation and cutting (extraction) of peatlands releases approximately 2 billion tonnes of CO_2 into the atmosphere every year. This represents around 10% of total global carbon emissions from all human activities. Emissions from Indonesia are especially high.

In the UK, only around 20% of peatlands are not degraded and remain in a pristine state; according to the government agency Natural England, the figure is just 1% for England. Rates of carbon sequestration in degraded peatlands are reduced; they may even become sources of carbon emission instead.

Some areas of UK peatlands have been *directly* affected by the practice of cutting (extracting) peat to burn as fuel, or for industrial use. For example:

- Peat has been used traditionally in many rural areas as a fuel source, both in the UK and other parts of the world. Once dried, it can be burned. Owing to the long time it takes to form, peat fuel can be viewed as a non-renewable resource, like coal. In Scotland and Ireland, peat is harvested on an industrial scale for use in power stations.

- Peat taken from raised bogs in lowland areas is commercially extracted by the horticultural industry and sold as nutrient-rich soil and compost in garden centres.

- Peat in Islay, Scotland, is burned as part of the whisky-making process. The Lagavulin and Laphroaig distilleries depend on peat smoke for the unique flavouring of their whisky. Peat fuel is also used to produce smoked fish and meat.

An even larger area of peatland has suffered *indirectly* as a result of land use changes and management practices (Table 11).

Table 11 Indirect damage to peatlands that can lead to reduced carbon storage

Drainage	■ Around one-quarter of English peatland is under cultivation ■ From 1640 onwards, large areas of wetland, such as the East Anglian fens, were drained for farming. This produced good agricultural land but degraded the peat ■ No longer waterlogged, the peat shrank, decomposed and became eroded by the wind. This, of course, released more CO_2 into the atmosphere
Pollution	■ Peatlands in Yorkshire have been subjected to decades of pollution from both Manchester and Sheffield ■ This pollution has led to widespread reduction in peat-forming plant species, making the peatlands more vulnerable to erosion
Burning	■ Large areas of peatlands throughout the UK are affected by moorland burning. This is a widespread land management practice in upland areas for the management of game, such as grouse ■ Burning the surface vegetation encourages the new growth of young heather which grouse feed on, but can damage the wet sphagnum mosses that create peat ■ Too much burning can remove the vegetation altogether, exposing the peat below, which can lead to subsequent rapid and widespread erosion of the peat during heavy rain (many upland areas receive high levels of orographic rainfall)

→

Knowledge check 11

Peatland areas play an important role in drainage basin water cycling. In a saturated state, peatlands generate overland flow quickly; can you suggest why?

Knowledge check 12

The production of peat-smoked meat and fish has helped some rural communities in the UK to diversify their economies in recent years. Can you establish synoptic links between your studies of Changing places and the Carbon cycle?

Table 11 *continued*

Grazing	■ Almost one-third of English peatlands now support invasive vegetation species not originally found there, meaning that the rate of peat formation — and carbon sequestration — may have slowed. These are often areas affected by livestock grazing
Forestry	■ Peatlands drained by the Forestry Commission (see p. 51) will begin to emit CO_2, and lose some soil carbon via leaching and erosion, but this may be offset over time by the CO_2 captured by the growing trees

Peatlands management and carbon store restoration

The majority of England's peatlands are currently sources of **greenhouse gases**, with notable emission 'hotspots' in the lowlands. Some upland peat areas are still capturing carbon, but most have become net sources for carbon emissions, as Table 12 shows. This is because of land use changes and mismanagement. Only undamaged areas — just 20% of all peatlands — continue to function as **carbon sinks**. All peatland areas that have been interfered with are now carbon sources, especially those used for agriculture.

Table 12 Carbon release and capture data for different peatland land uses (source: Natural England)

Land use (in peatland areas)	Net carbon flux (tonnes of CO_2 per hectare per year)
Cultivated and temporary grass	22.4 net loss
Improved grassland	8.7 net loss
Rotationally burnt	2.6 net loss
Afforested	2.5 net loss
Bare peat	0.1 net loss
Overgrazed	0.1 net loss
Undamaged	4.1 net storage

The rationale for peatland restoration

Protecting and restoring peatlands is important for maintaining biodiversity. Peatland restoration could also help the UK and other countries meet their reduction targets for long-term greenhouse gas emissions. By restoring peat environments, local and national governments may be able to offset greenhouse gas emissions produced by economic activity.

This is an **ecosystem services** or 'environmental economics' approach to nature conservation. In the long term, it is hoped a strong economic case will be made for widespread restoration of peat if a functioning global **carbon market** can be developed as part of climate change mitigation efforts.

Restoration in practice

Restoration efforts usually include:

■ the re-establishment of a plant cover dominated by peatland species including sphagnum mosses
■ the 're-wetting' of drained peatlands by raising and stabilising the local water table

Greenhouse gases are those atmospheric gases that absorb infrared radiation and cause world temperatures to be warmer than they would otherwise be. They include carbon dioxide (CO_2) and methane (CH_4).

A **carbon sink** is an environment or organism that absorbs more carbon than it releases as carbon dioxide, resulting in net carbon storage.

Self-study task 15

Analyse the variations shown in net greenhouse gas flux data in Table 12. Suggest reasons for these variations.

Ecosystem services are benefits for humans provided by ecosystems that would create economic costs if they became unavailable. Examples of ecosystem services include products such as food and water, regulation of floods and carbon storage.

A **carbon market** is a proposed way of using market forces to bring about carbon emissions reductions. Activities which help to store carbon — such as peatland restoration — might be rewarded with credits that can be offset against activities that produce CO_2.

CO_2 emissions from the decomposition of the peat are reduced immediately after a previously drained peatland is 're-wetted' and anaerobic conditions have returned. In the longer term, it is hoped that new peat formation will begin to occur, with ongoing sequestration of atmospheric carbon. The environment will be restored as a carbon sink (rather than a carbon source).

A good example of ongoing restoration is the upland peatlands of the southern Pennines in Yorkshire. These highly degraded peatlands were extensively drained in the past and have been severely eroded by overland flow and **gullying**. In recent years, actions have been taken to:

- block erosional gullies with stone dams to help raise the water table and retain moisture in the peatlands, thereby restoring anaerobic conditions
- reintroduce wetland species, for example by re-seeding of the landscape to restore its original vegetation cover, including cottongrass and sphagnum moss
- apply seed and fertiliser using helicopters

It is too early to judge whether these actions have been entirely successful but initial signs suggest the re-vegetated surfaces are much less prone to erosion (Figure 34).

In Scotland, larger areas of peatland that were drained for forestry are being restored to their natural state. Drains have been blocked, and trees removed from the landscape. So far, these techniques appear effective in reducing carbon loss from the peat.

Gullying is the process whereby gullies are formed on a land surface by heavy rainfall. Overland flow becomes concentrated in shallow channels which then combine to form deep gullies that dissect the land surface.

Figure 34 Peatland restoration

Summary

- The process of peat formation has taken place over thousands of years. Peatlands are located in poorly drained sites throughout the UK, and their characteristics vary according to relief and altitude.
- Undisturbed peatlands serve as a carbon sink, and are important for carbon storage at a local and global scale.
- Agriculture, drainage, land use changes and peat cutting have led to the degradation of the majority of UK peatlands. Most no longer function effectively as carbon stores and have become carbon sources instead.
- Peatland restoration in the UK and elsewhere can be a valuable way of reducing net carbon emissions.
- Management strategies for peatland restoration include re-wetting and the re-introduction of wetland species.

■ Links between the water and carbon cycles

Increasing atmospheric carbon storage

Taken together, a range of evidence suggests that Earth's climate is currently warming and changing; the US National Oceanic and Atmospheric Administration (NOAA) says the signs are 'unmistakable'.

- In 2015, global mean surface temperature (GMST) reached a new record high of +0.87°C relative to the 1951–80 average GMST.
- The ten warmest years since 1880 have all been since 1998.

The prevailing viewpoint held by the large majority of the world's climate scientists is that global warming is caused by increased greenhouse gas (GHG) emissions, and that humans are the cause of this. The UK Meteorological Office says that the signs of warming 'have human fingerprints' all over them. In the fifth climate change assessment (2013), Intergovernmental Panel on Climate Change (IPCC) scientists reported they were 'virtually certain' that humans are to blame for 'unequivocal' global warming.

The key steps in the argument are as follows:

- Carbon dioxide emissions have been rising since 1750, the start of Europe's Industrial Revolution, from a level of 280 ppm (parts per million) to 406 ppm at the start of 2017. This represents an increase of around 45%. Even more worryingly, if you convert other GHGs (methane, nitrous oxide) into their equivalent amounts of CO_2, then you find that we have reached a level in excess of 470 ppm of CO_2 equivalents.
- This is because population growth and economic development have led to: worldwide use of carbon-rich fossil fuels as an energy source; widespread deforestation; cement manufacturing; enhanced methane emissions derived from livestock (bovine flatulence) and the decomposition of organic wastes in landfill sites.
- A so-called 'hockey stick' trend line can be seen in carbon dioxide data: there was gradual growth up to the early 1900s and then very steep growth. Measurements have been taken offshore at Mauna Loa every year since the 1950s and these show a rise each year of around 1–2 ppm. Older evidence comes from 'fossil air' trapped in ice. Permanent ice in high mountains and polar ice caps has built up from snow falling over hundreds of thousands of years. Cores between 3 km and 4 km long have been extracted from the Vostok ice sheet in Antarctica, producing ice that is more than half a million years old and containing bubbles of ancient air. After analysing the evidence, scientists believe that carbon dioxide and methane concentrations are now higher than at any time in the last 800,000 years (Figure 35).

Knowledge check 13

Carbon dioxide and methane are the two carbon-based greenhouse gases. Which human activities do you associate with increased emissions of which gas?

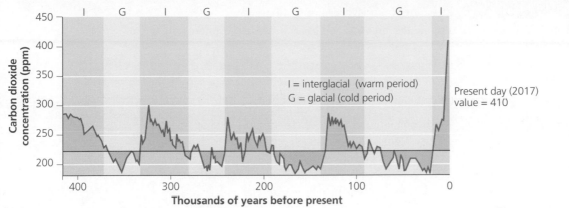

Figure 35 Atmospheric carbon dioxide concentrations measured from Vostok ice cores and recent Mauna Loa data

Increasing carbon emissions and the energy budget

Rising GMST is believed to be a product primarily of increases in the atmospheric carbon store which have impacted on the Earth's **energy budget** (Figure 36).

The **energy budget** is the state of balance between incoming solar radiation received by the atmosphere and the Earth, and the re-radiated heat or reflected energy.

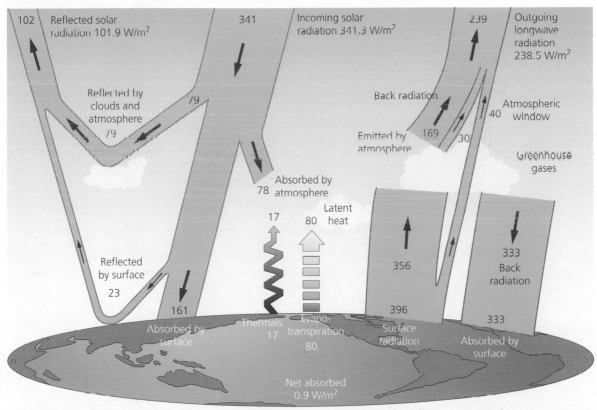

The global annual mean Earth energy budget for the March 2000–May 2004 period (W/m²). The broad arrows indicate the schematic flow of energy in proportion to their importance.

Figure 36 Global energy budget (source: Trenberth, 2009)

The Earth's climate is driven by incoming shortwave solar radiation:

- Approximately 31% is reflected by clouds, aerosols and gases in the atmosphere and by the land surface.
- The remaining 69% is absorbed; almost 50% is absorbed at the Earth's surface, especially by oceans.
- 69% of this surface absorption is re-radiated as longwave radiation.

A large proportion of the longwave radiation emitted by the surface is absorbed by the atmosphere (clouds and greenhouse gases), from where it is re-radiated both back to the Earth's surface and to space. This 'trapping' of longwave radiation in the atmosphere is what gives a life-supporting average of 15°C, the 'natural greenhouse effect'. Without this process, the Earth would be −18°C (too cold for life to have evolved).

To summarise, increasing carbon emissions (carbon dioxide and methane) mean that more heat is being radiated back towards the ground surface (this is the climatic equivalent of putting extra blankets on a bed). The energy budget is changing; more heat is being retained, resulting in a warmer, more energetic climate system. However, there is great uncertainty even among experts over the extent and timing of future climate forecasts.

Impacts of rising greenhouse gas emissions on the water cycle

There are signs that the world's water cycle and oceans have already been affected by recent increases in the atmospheric carbon store. Pages 12–13 explore changes in cryospheric storage (major ice sheets are losing mass, many land-based glaciers are shrinking and Arctic sea-ice cover has reduced significantly since 1979). Table 13 shows additional impacts of increased atmospheric carbon storage (and the rise in GMST this brings) on water, weather and oceans.

An **extreme weather event** is the occurrence of a value of a weather or climate variable (such as precipitation, wind strength or temperature) that is above or below a threshold value near to the previously observed maximum or minimum value.

Table 13 Impacts on water, weather and ocean of increased atmospheric carbon storage

Amount, type and patterns of precipitation (including extreme weather events)	■ In a warmer world, more evaporation takes place over the oceans — and what goes up must ultimately come down. Climate scientists believe rainfall patterns are changing in many parts of the world as the world's oceans warm (see p. 14)
	■ UK annual average rainfall has not changed since the eighteenth century. However, in the last 30 years more winter rainfall has fallen in heavy events
	■ Climate change predictions for the UK suggest that the total amount of precipitation may not change but that the pattern will become more seasonal. There could be an increased frequency of frontal rainfall in winter; and a greater probability of summer drought because of reduced frontal rainfall (see p. 28)
	■ Extreme high intensity rainfall in 2007 led to pluvial (surface water) flooding in some British cities, which caused £3 billion damage; high intensity rainfall collected in low-relief areas where homes were located (often in places far from a river or coastline). Higher temperatures in the future could mean an increased probability of high intensity rainfall events (Figure 37)
	■ Some scientists believe the probability of extreme weather events occurring has 'significantly' increased. The 2012 *State of the Climate* report by the American Meteorological Society (AMS) showed a new emerging consensus among leading meteorologists
River discharge	■ A consensus has emerged among UK meteorologists that the UK has entered a flood-rich period marked by more intense winter rainfall events. This is in line with IPCC climate change projections. Since the 1980s, river flooding has increased in size and duration during winter

	■ In contrast, lower river discharges may become more common in southern England as a result of less frontal rainfall in summer ■ Although high intensity convectional rain can be expected sometimes in summer, it is important to remember that it can lead to infiltration-excess overland flow. This means that depleted soil water and groundwater stores are not recharged. Under these conditions, river discharge will quickly return to a low level after a brief flashy hydrograph response to the rainfall
Sea level rise	■ Historically, sea levels have always been higher during warm periods and lower during cold phases. In the future, a warmer climate is predicted to bring a positive (rising) eustatic sea level change for two reasons. Together, thermal expansion of the oceans (see p. 14) and worldwide glacial melting are projected to bring a series of eustatic sea level rises ■ Already these combined processes are giving rise to a total global average sea level rise of approximately 3 mm per year. In total, global sea level has risen by 200 mm since 1900 and the IPCC projects a further world sea level rise of 260–820 mm by 2100, mostly due to thermal expansion ■ Significant glacier and permafrost meltwater runoff will produce another metre of sea level rise by around 2200 (even if carbon emissions are cut significantly) ■ Without any action to cut emissions, the complete loss of the Greenland and Antarctica ice sheets would result ultimately in a eustatic rise of nearly 70 m (although not for hundreds or even thousands of years yet)
Acidification of the oceans	■ It is estimated that nearly one-third of the CO_2 that has been released into the atmosphere by human activity has diffused into the ocean. If this had not occurred, then atmospheric concentrations of carbon dioxide would be even higher than they are ■ Dissolving carbon dioxide in the ocean creates carbonic acid, however. Since 1750, the pH of the ocean's surface has dropped by 0.1, a 30% change in acidity ■ The increased acidity of ocean water has harmful effects for coral reef organisms and other sensitive species. Carbonic acid reacts with carbonate ions in the water to form bicarbonate. However, those same carbonate ions are what coral need to create their calcium carbonate shells. With less carbonate available, the animals need to expend more energy to build these shells. As a result, the shells end up being thinner ■ Increasing acidity could have a range of unknown but potentially harmful effects on different marine species, including impacts on metabolic rates, reproduction and immune systems

A **eustatic** sea level change is a worldwide rise or fall in average sea level, resulting from a warming or cooling climate affecting the volume and/or depth of water in the oceans.

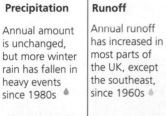

	Precipitation	Runoff	River flow	Evaporation
What's already happened?	Annual amount is unchanged, but more winter rain has fallen in heavy events since 1980s ●	Annual runoff has increased in most parts of the UK, except the southeast, since 1960s ◔	There has been increased frequency and magnitude of winter flooding since 1980s ◔	There is little reliable evidence to show how evaporation and transpiration have changed ◔
What may happen in the future?	Precipitation will become even more seasonal, but annual total will stay about the same ◔	Predictions are hard to make because of uncertainty over both rainfall and evaporation ◔	Some studies project slightly increased flows and increased floods, particularly in winter ◔	Potential evaporation will increase due to higher air temperatures this century ◔

● High confidence this is linked with climate change ◔ Lower confidence

Figure 37 Recorded and projected changes in hydrological flows for the UK

Links between the water and carbon cycles at the local scale

Throughout this book, significant interactions between water cycling and carbon cycling have been highlighted. This has included consideration of how:

- the two cycles interact directly when carbon is transported in solution by river water (see p. 44)
- ecosystems function as stores of both water and carbon, and influence the way a range of carbon transfers and water flows operate (see pp. 40–44)
- deforestation can lead to increased overland flow, which in turn causes soil erosion and the permanent loss of carbon storage capacity (see p. 51)
- increasing atmospheric carbon concentrations are changing global climate and water cycle flows and stores (see p. 58)

Additional integrated case studies of land use can be used to illustrate links between water and carbon cycling at the local scale. For example, the process of **desertification** serves to emphasise:

1 how water and carbon cycles are closely coupled

2 how water cycle changes are linked with soil and vegetation loss

3 the interdependence of soil and vegetation health

Figure 38 shows how desertification in a developing world region such as the Sahel is driven partly by climate change (or fluctuations) and also by land use change.

- Reduced vegetation cover may lead to reduced carbon sequestration in biomass (due to lower photosynthetic fixation of carbon) and also reduced soil carbon.
- Less vegetation cover may reduce the remaining soil's infiltration capacity (the soil surface may become crusted or compacted); this means infiltration-excess runoff will take place when rain does occur (see p. 18), resulting in soil erosion and gullying.
- Accelerated soil erosion will reduce soil carbon storage even more.
- With less soil cover, ecosystem net primary productivity (NPP) will fall further.

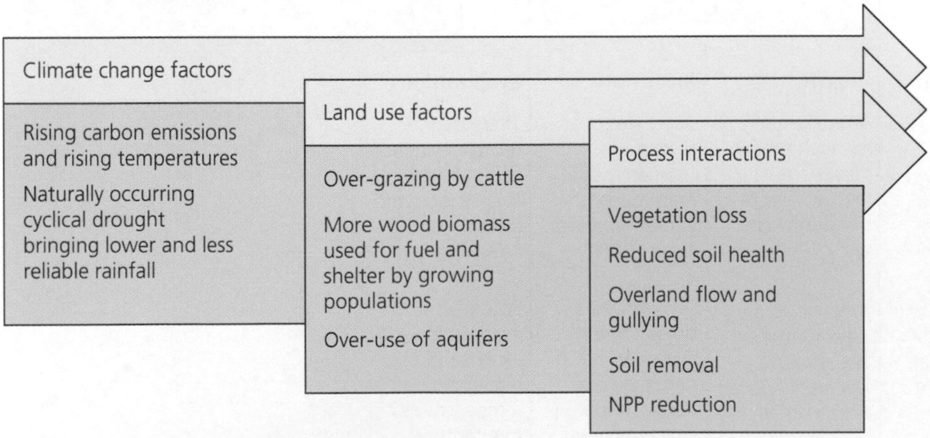

Desertification is the process by which land becomes drier and degraded as a result of climate change or human activities, or both.

Climate change factors		
Rising carbon emissions and rising temperatures	**Land use factors**	
Naturally occurring cyclical drought bringing lower and less reliable rainfall	Over-grazing by cattle	**Process interactions**
	More wood biomass used for fuel and shelter by growing populations	Vegetation loss
		Reduced soil health
	Over-use of aquifers	Overland flow and gullying
		Soil removal
		NPP reduction

Figure 38 How water and carbon cycle changes are linked when desertification is taking place

Fieldwork

You could investigate links between the water and carbon cycles at the local scale, for instance by estimating how carbon is being moved in solution by a stream.

Studying cycle linkages in the UK

Significant interactions between water cycling and carbon cycling can be explored at the local scale (Figure 39). Floodplains, parks and even small gardens or patches of wasteland can provide a study context. For example, in autumn and winter you can observe fallen leaves being washed by heavy rainfall into gutters and sewers. In rural and urban drainage basins alike, high levels of overland flow and channel flow in the water cycle are responsible for transporting a large volume of carbon (stored as leaf litter biomass and soil organic matter) away from the land.

Figure 39 Forest floor litter in Epping Forest

Self-study task 16

Figure 39 shows organic litter in Epping Forest.
- What role will precipitation play in helping this litter to decompose?
- What role could overland flow play in transporting this litter to other places, or into a stream?
- In the longer term, what role will this litter play in the water cycle if it remains here, decomposes, and becomes stored as humus in the soil?

Summary

- Based on the post-1750 carbon dioxide trend, most climate scientists agree that humans are responsible for increasing atmospheric carbon storage.
- Increasing carbon emissions have affected the Earth's changing energy budget, resulting in a rise in global mean surface temperature (GMST).
- Increased atmospheric carbon is starting to affect rainfall and water cycle flows in complex ways, and the UK is expected to experience wetter winters and drier summers in the future.
- Climate change projections suggest the oceans will become more acidic and global sea level will keep rising.
- Local-scale links between the water and carbon cycles can be explored through (1) the study of desertification in parts of the Sahel, and (2) the seasonal removal of leaf litter in the UK, including urban areas.

■ Feedback in and between the carbon and water cycles

Equilibrium, feedback and thresholds in natural systems

References to the concept of equilibrium have recurred throughout this book's analysis of the global water and carbon cycles. Most natural systems, unaffected by human activity, exist in a steady-state equilibrium. As explained on p. 12, this means they have highly variable inputs, outputs, flows and levels of storage in the short term, but in the long term a broad state of balance is maintained. Another specialised concept which helps us to model system dynamics over time is **feedback**.

Feedback is the automatic response of a system when a change to the flow of energy or materials occurs.

- **Negative feedback** occurs when a system adjusts itself in ways that lessen or cancel out the effect of an initial disruption that has interfered with the system's normal operation. The disruption to the system triggers changes in other system elements which act in *the opposite direction* to the initial change. As a result, equilibrium or balance is restored. Put simply, natural systems have their own internal checks which help 'put the brakes on' and restore balance.

- **Positive feedback** loops, in contrast, are knock-on effects in natural systems that act to *accelerate and amplify* any changes that have already started to occur after a disruption occurs. When one element of a system changes, it upsets the overall equilibrium, or state of balance, thereby leading to changes in other elements which reinforce what is happening. Put simply, the system starts to 'spin out of control'.

Figure 40 shows examples of positive and negative feedback that help illustrate how these processes operate. The two models show possible scenarios that may result from more carbon dioxide being added to the atmosphere. The first scenario suggests accelerated warming as various system changes take place that strengthen one another. In contrast, the second scenario of negative feedback involves some 'knock-on' effects cancelling out the impact of the initial change (i.e. a temperature rise triggered by an increase in carbon dioxide).

There is widespread concern among the world's climate scientists that positive feedback effects associated with human emission of carbon dioxide will be far greater than negative feedback effects in the coming decades. As a result, there is a significant risk of a high rise in GMST of 4–6°C occurring by 2100 (see p. 69).

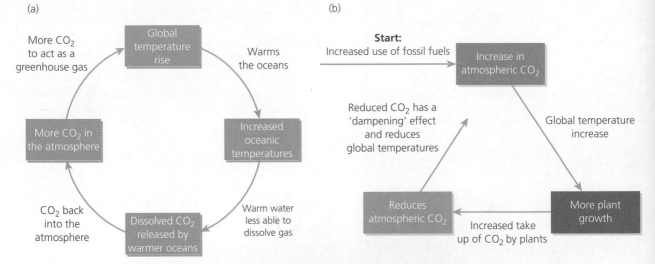

Figure 40 Contrasting effects of (a) positive feedback and (b) negative feedback in the climate system

The system threshold

In studies of feedback, the concept of a **system threshold** can be important. This is a critical limit or level that must not be crossed if accelerated and potentially irreversible changes are to be avoided, making it much less easy for the previous equilibrium state to be restored. It is possible to envisage a significant change to either a water or carbon system forcing it across a threshold from which the previous equilibrium state cannot be regained (Figure 41). For example:

■ Many scientists believe that a global temperature rise in excess of 2°C will lead to widespread and irreversible changes in the global carbon cycle because of positive feedback processes.

■ At a local level, overgrazing of vegetation in an area at risk of desertification could cross a threshold: if too much vegetation is lost, widespread soil erosion could follow. Once this happens, vegetation cannot easily re-colonise. Irreversible loss of both the vegetation and soil store means that local water and carbon cycles will be permanently changed.

Self-study task 17

The concepts of equilibrium, feedback and thresholds feature in many other geography topics. These concepts are widely used in the study of coastal and glacial landscapes. For instance, a model of a beach sediment store may show how it exists in a state of equilibrium, with balanced inputs and outputs of sediment moved by the process of longshore drift. Place studies can make use of the same concepts too. In human geography, the decline of a rural settlement may cross a threshold (a 'point of no return') if a vitally important service closes, such as the local primary school (beyond this point, the settlement will no longer have a sustainable future). Have equilibrium, feedback and thresholds featured anywhere else in your course?

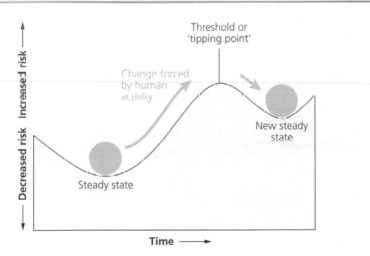

Figure 41 The threshold concept

Feedback impacts in and between the water and carbon cycles

Several specific illustrations follow which show how feedback effects can take place in and between the water and carbon cycles. In some cases, negative feedback helps balance to be restored. In other cases, positive feedback may accelerate the rate of change beyond a 'tipping point' of no return.

Cryosphere feedback

During periods when the Earth has grown warmer in the past, temperature changes may have accelerated because of the loss of ice cover. Ice has a high **albedo** of around 80%, which means it reflects four-fifths of all incoming solar radiation. If white-coloured sea ice melts in the Arctic, for instance, darker-coloured water will be revealed. The water has a lower albedo than ice because of its darker colour. It therefore absorbs more heat, warms up and is likely to melt even more sea ice, thereby opening up more areas of open ocean, and so on (Figure 42). As the water warms up, so too do the air masses in contact with it. The result is accelerated warming of the atmosphere.

> **Albedo** is the fraction of solar radiation that a surface reflects. White surfaces have the highest albedo, or reflectivity.

This positive feedback process means that even a small change in sea ice coverage has a potentially significant impact on global climate. In theory, even a small reduction in sea ice cover could lead ultimately to an ice-free Arctic.

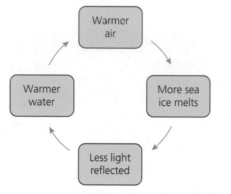

Figure 42 A simple theoretical positive feedback loop which could accelerate cryosphere shrinkage

Feedback cycles are not always predictable, however. Consider how Figure 43 suggests an alternative scenario for the Arctic.

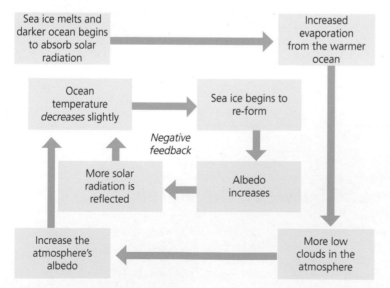

Figure 43 How a negative feedback loop could nullify cryosphere shrinkage

1 Ice melts to expose the darker ocean, which absorbs more sunlight.

2 As the ocean warms, there are increased levels of evaporation.

3 This creates more cloud, especially in the lower atmosphere.

4 The light-coloured cloud has a high albedo and reflects incoming solar radiation (much as the ice used to).

5 As a result, less light reaches and is absorbed by the ocean surface.

6 The temperature of the water falls, as does that of the air mass in contact with it.

This is an example of how negative feedback could help the climate system to self-correct if Arctic ice begins to melt.

Methane feedback

Methane is a powerful greenhouse gas, enormous volumes of which are stored in frozen soils in Earth's **permafrost** regions. Around one-quarter of the Earth's surface is affected by continuous or sporadic permafrost, including tundra, polar and mountain regions. Globally, permafrost covers 23 million square kilometres (mostly in Earth's northern hemisphere). It formed during past cold glacial periods and has persisted through warmer interglacial periods, including the Holocene (the last 10,000 years).

Reports indicate that methane-storing permafrost is now shrinking at an alarming rate. According to scientists at NASA, temperatures in Newtok, Alaska, have risen by 4°C since the 1960s, and by as much as 10°C in winter months. The effects of permafrost melting are potentially magnified over time by positive feedback loops.

1 As the atmosphere warms, more permafrost is expected to melt.

2 This will release large amounts of methane (some researchers estimate that the volume of this gas stored in permafrost equates to more than double the amount of carbon currently in the atmosphere).

3 The atmosphere will warm up even more quickly.

4 Even more methane will be released by more melting permafrost, etc.

In theory, this could easily take Earth's climate beyond a 'tipping point' threshold (the point of no return).

Terrestrial and marine carbon feedback

Figure 44 shows yet another positive feedback model. Here, combined terrestrial and marine feedback loops are shown which both result from, and accelerate further, global temperature rise. The elements of this model include:

- increasing water vapour in the atmosphere; because water vapour is also a greenhouse gas, this could lead to further temperature rises (however, huge scientific uncertainty exists in defining the extent and importance of this feedback loop: as water vapour increases in the atmosphere, more of it will eventually also condense into clouds, which are more able to reflect incoming solar radiation)
- terrestrial permafrost melting and methane release
- a reduction in seawater's ability to absorb surplus CO_2 from the atmosphere: warmer water is less effective at absorbing CO_2 than colder water, and so may begin to release, rather than absorb, the gas if temperatures continue to rise

Permafrost is ground (soil or rock and included ice) that remains at or below 0°C for at least two consecutive years. The thickness of permafrost varies from less than 1 m to more than 1.5 km.

There are two further terrestrial and marine feedback loops to consider (which are not shown in Figure 43):

- ocean acidification could impact negatively on coral and marine ecosystem health in ways that reduce biological sequestration of CO_2 in the oceans, i.e. the biological pump might become less effective (see p. 43)
- the global biome pattern may change in response to rising GMST. If the **tree line** moves north — and the coniferous tree biome grows in size — then more carbon might be stored. But if large areas of grassland are damaged by desertification, carbon storage will be lost. Overall, there is huge uncertainty over what the net effect of higher temperature rises will be on carbon storage patterns linked with biome distributions

The **tree line** is the boundary between the coniferous forest and tundra biomes.

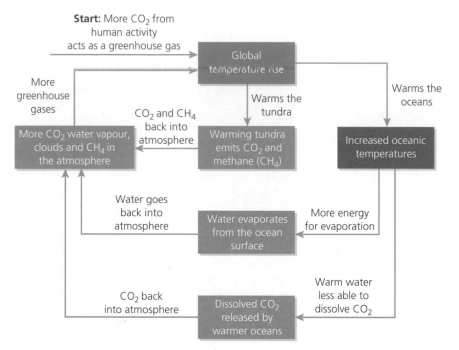

Figure 44 A system diagram showing how interrelations between climate change, water flows and carbon flows could create positive feedback loops

The implications of system feedback for life on Earth

Figure 45 shows two different climate change pathways modelled by the IPCC. As you can see, in each case a range of temperature rises is shown, with differing implications for life on Earth. Having read about the complexity of positive and negative feedback loops, you should by now have a good understanding of why such a high level of uncertainty exists, even among scientific experts. The GMST rise prediction for a high emissions scenario varies from 3°C to 5°C. Among many other things, this reflects uncertainty around the strength and timing of possible positive feedback loops linked with Arctic permafrost thawing.

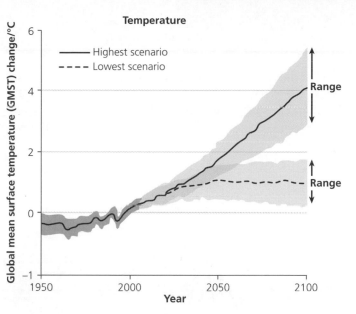

Figure 45 Two future pathways modelled by the IPCC

Self-study task 18

Study the different projections and estimates shown in Figure 45. Describe the trends shown. Suggest why is there still so much uncertainty about what will happen to global mean surface temperature.

There are also serious implications linked with the potential disappearance of large parts of the cryosphere (see p. 10), the meltwater of which currently feeds large rivers upon which hundreds of millions of people depend for their water security. For example, China is home to 1.3 billion people, many of whom have recently moved to urban areas. By 2025, there will be 220 Chinese cities with populations in excess of 1 million, eight of which will be **megacities** (settlements with over 10 million residents). However, the sustainability of these growing settlements is threatened by a new lack of water security for the southeast Asian region as a whole.

- Many major rivers are fed by seasonal meltwater runoff from major glaciers in the region, notably the Himalayan Plateau.
- Every summer, ice melts and feeds Asia's largest rivers. Fresh snowfall each winter replenishes the glaciers, meaning that over time the meltwater cycle is sustainable.
- However, climate change system feedback threatens to permanently reduce the size of glacial ice stores in the region. Although this will increase meltwater in the short term, in the long term it could lead to dangerous water shortages because there will be very little ice left to melt!

A wicked problem

This chapter has demonstrated that climate change science is far from simple. There is uncertainty over the operation of feedback loops. Nor can we predict with any certainty what kinds of economic, demographic, technological and political changes will take place globally, or in particular countries, that may influence the success or failure of climate change mitigation efforts. As a result, climate change presents us with a **wicked problem** which — because of its complexity — defies attempts to

A **wicked problem** is a challenge that cannot be dealt with easily owing to its scale and/or complexity. Wicked problems arise from the interactions of many different places, people, things, ideas and perspectives within complex and interconnected systems.

establish clearly what its exact effects are likely to be! Unfortunately, this reasonable uncertainty is seized upon by climate change sceptics as a reason to avoid taking any action to reduce carbon dioxide and other GHG emissions.

Figure 46 provides a useful summary of some of the many different influences on climate change processes and the high levels of complexity and uncertainty which accompany any attempt to model the implications of rising carbon emissions for life on Earth.

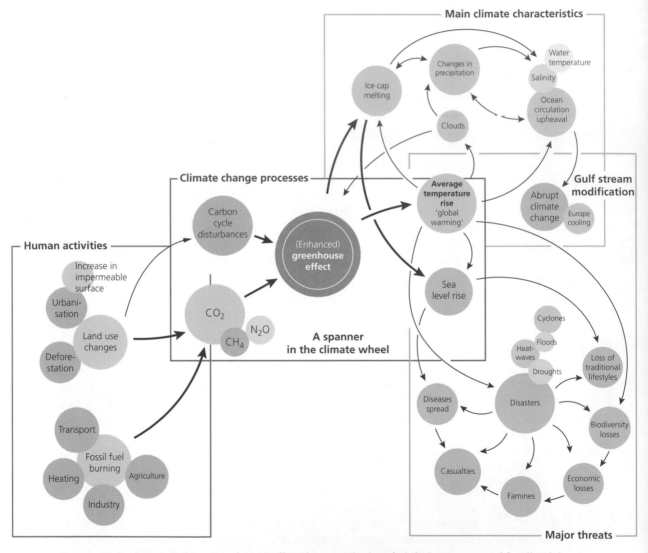

Figure 46 A wicked problem: understanding the complexity of global systems and feedback loops affecting climate change (source: Cameron Dunn)

Summary

- Negative feedback loops help to restore equilibrium in natural systems after a disturbance or change.
- In contrast, positive feedback can result in a permanent change of state for a system. If a critical level, or threshold, is crossed then it may become impossible for changes to be reversed.
- There are signs that the cryosphere is beginning to shrink because of climate change. This shrinkage may accelerate because of positive feedback linked with melting ice and the changing albedo of land and water surfaces.
- Permafrost melting releases methane, a powerful greenhouse gas. The rate of melting could accelerate because of powerful positive feedback effects in the future.
- There are many more potential feedback loops for water cycling and carbon cycling in both terrestrial and marine environments. As a result, climate change modelling is subject to a greater degree of uncertainty.
- Climate change is the ultimate 'wicked problem' on account of the sheer scale of its challenge and the complexity of the process interactions and feedback loops associated with global temperature rise. As a result, the implications for life on Earth remain hard to predict with any level of certainty.

Fieldwork and investigative skills

■ Planning an independent investigation

As part of your full A-level geography course, you must complete a fieldwork-based **independent investigation** (also known as the non-examined assessment, or NEA).

- The single independent investigation consists of a written report.
- It is worth 80 marks, contributes to 20% of the A-level qualification and is marked by your teacher.
- Ideally, it should be between 3,000 and 4,000 words in length.
- The focus must be related recognisably to the A-level geography specification.

The independent investigation provides a great opportunity for you to devote time to the in-depth study of a topic that really interests you, allowing you to acquire and develop skills that have lifelong value. However, it also presents a major challenge. You will be responsible for the delivery of a substantial piece of high-quality work, but with far less teacher input and guidance than you may have been used to in the past.

Your investigation *must* make use of a significant amount of **primary data** requiring out-of-classroom fieldwork.

- You are, in any case, required to undertake four teacher-assisted days of group fieldwork during your A-level course; this must cover topics in both human and physical geography.
- Some students will attempt to collect all the necessary primary data for their independent investigation during these four days of group fieldwork; other students may make their own arrangements to carry out fieldwork alone (and perhaps on a different topic) at another time, probably during the school holidays.

In summary, your independent investigation may incorporate field data collected *either* while working entirely independently (though not necessarily alone) in your own time *or* as part of your four compulsory days of teacher-assisted group fieldwork.

Primary data consist of brand new first-hand information which you have collected by yourself or working in a group.

If you choose the latter route, however, there are important guidelines you need to follow in order to demonstrate *independence* (see p. 73).

The route to enquiry

Table 14 and Figure 47 show the six stages of enquiry for your independent investigation. Make sure there is no doubt in your mind about what will be involved at each stage. It is advisable to pay close attention to the mark allocation from the outset because this will guide you in knowing how long to spend on each stage of the enquiry process.

Table 14 The WJEC/Eduqas model of enquiry (adapted from the specifications)

	Stage	Description	Marks
1	Context setting	■ Begins with a 250-word **abstract** (or synopsis) of the investigation which clearly states its relevance and what you hope to achieve ■ Identifies background to the topic focus, including any conceptual framework or important theoretical background such as a model (e.g. the carbon cycle). Ideally, the models and theories used should be **contemporary** ■ Gives consideration to any **risks** (the risk assessment) and **ethical issues** arising from the delivery of the independent investigation ■ Outlines the case study area to provide context	10
2	Field investigation methods	■ Uses a range of appropriate fieldwork methods to observe, measure and record geographical phenomena ■ Demonstrates critical thinking by **justifying** the methods used (such as their viability and appropriateness for the research question(s) under investigation) ■ Outlines **own independent contribution** to the design of the methodology	15
3	Data presentation techniques	■ Communicates **primary** and **secondary** data findings using a range of appropriate presentation techniques, most likely including spatial (mapping), graphical (charts) and representational (photographic and other visual) methods ■ May comment on the **appropriateness** of the techniques used, and the extent to which they facilitate suitable analysis of quantitative and qualitative data	10
4	Analysis and interpretation	■ Analyses the data and information collected, most likely making use of statistical methods and measurements ■ Interprets these data and any statistical findings, most likely by identifying patterns, trends, anomalies and causal relationships ■ **Justifies** any evidence-based arguments derived from data analysis and interpretation, most likely by reflecting on the **significance** of the investigation's findings, and levels of **confidence** in what the data show	15
5	Conclusions and presentation	■ Concludes in a detailed and well-evidenced way which draws on **all** of the investigation's findings ■ Refers back to the theoretical or conceptual underpinning of the research, and the extent to which the study's findings correspond with established geographical knowledge or perhaps offer new or novel insights ■ Follows presentation rules well (e.g. the referencing of secondary data)	10
6	Evaluation	■ Provides a succinct reflection on every stage of the investigation and the extent to which strengths and limitations can be identified in what was done ■ Offers a comprehensive overview of primary and secondary fieldwork methods and data, in terms of their **accuracy**, **reliability** and possible **bias**, and so reflects on the extent to which the conclusions can be relied upon ■ Reflects on how the views and interests of different stakeholders, if relevant, may have influenced the investigation's findings; considers any **ethical** dimensions ■ Suggests ways in which a similar investigation could be improved upon or extended with further research ■ Provides a comprehensive **bibliography** and relevant appendices	20

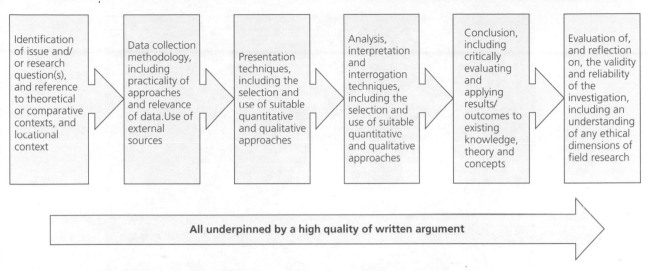

Figure 47 The independent investigation pathway

Finding a research topic focus

Coming up with your own geographical idea or focus can be challenging, especially if you were not given many opportunities for independent learning at GCSE. This might be the first time you have taken on a substantial assignment where such a greater burden has been placed upon you to come up with all the questions and answers yourself. Perhaps the most important piece of advice is to make sure you are investigating something you are genuinely interested in finding out more about, because this will hopefully motivate you to get the work done.

If you are visiting a field studies centre and your teacher has suggested a **topic frame** and case study area already — such as the study of coasts or places at a site near the centre — it is vital that you can identify a particular narrower aspect of this topic which interests you enough to want to spend a significant amount of time researching and writing about it.

■ A good way to approach this might be to spend time looking carefully at photographs and maps of the place you will be visiting.

■ What interests you, or does not interest you, about maps, aerial photographs or Google Earth views of the chosen environment?

■ Which questions enter your head initially when you look at photographs of the case study area? It is quite likely that the things which interest you most will spring to mind first.

Figure 48 shows a photograph of a woodland environment (Epping Forest in Essex) which a Year 12 geography class have been told they will be visiting next term. The questions at the margins of the photograph all come from different class members who have been shown the photograph prior to their visit. Any of these questions might form the basis for a good independent investigation. As you can see, some are more science-based whereas others are focused on management interactions. Which question would you be happiest to adopt, and why?

The **topic frame** is the broad topic area, such as Changing places, or Coastal landscapes, to which your independent investigation belongs.

How much water infiltrates into the ground when it rains?

How much carbon do the oldest trees here store?

Is this woodland managed sustainably?

How much seasonal variation is there in interception storage?

How resilient is Epping Forest to extreme weather damage?

What are the main risks Epping Forest faces?

How is Epping Forest represented as a place in the media?

How much seasonal variation is there in carbon storage?

Figure 48 Geographical questions asked by a class about a picture of the woodland environment they will be visiting, any of which could form the basis for an independent investigation

On the other hand, if you have the freedom and confidence to work entirely independently, then you can (1) devise an independent investigation that is derived from any part of the A-level geography specification, and (2) select your own case study area.

■ Figure 49 shows possible topic areas you could draw from.

■ Table 15 shows planning-stage questions that are frequently asked by students when considering whether to 'think outside the box' or settle to work within a more restricted topic frame that their teacher has suggested (such as Changing places).

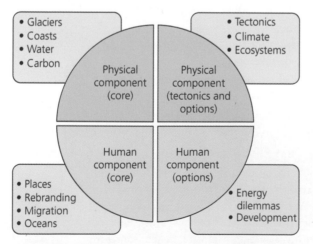

- Glaciers
- Coasts
- Water
- Carbon

Physical component (core)

- Tectonics
- Climate
- Ecosystems

Physical component (tectonics and options)

Human component (core)

- Places
- Rebranding
- Migration
- Oceans

Human component (options)

- Energy dilemmas
- Development

Figure 49 Possible investigation topic areas drawn from the WJEC and Eduqas A-level specifications

Table 15 Frequently asked questions for the planning stage

Can I ...		Explanation
Choose any topic I want?	✔	■ In theory, yes ■ In practice, your teacher may have a strong preference for the entire class to work within the same broad topic frame, such as the Water cycle or Changing places
Wait until the fieldtrip has begun before choosing a title?	✘	■ This is a risky thing to do ■ You will have little time to reflect on the choice you make, or to carefully refine your aims and objectives
Gain credit from using data that the entire class collected during our fieldtrip?	✔	■ It depends. You may be able to gain credit for using shared data if you can demonstrate that *you played an active role in developing some of the methods that were used*. Did you choose the sampling strategy, for instance, or write some of the questions used on the survey?
	✘	■ Any data collected prior to you developing your own investigation title and methodology cannot be treated as primary data and must instead be classed as secondary data
Carry out research anywhere?	✔	■ In theory, yes ■ In practice, you need to think about the accessibility, cost and risks associated with the use of particular sites
Choose my own title?	✔	■ Yes. In fact, you *must*

Table 16 provides a useful summary of what is expected of you when it comes to working independently or collaborating with other students.

Table 16 Working in a group, working alone, or both?

Group fieldtrip-derived independent investigation	'Solo mission' independent investigation	Combined approach independent investigation
■ You must complete 4 days of fieldwork as part of your A-level course. This fieldwork visit (or 'fieldtrip') could serve as the basis for your own independent investigation ■ But you MUST demonstrate that you **personally** had an active role in designing the fieldwork: you cannot just 'follow orders' ■ This means you need to plan ahead carefully. *Well before the fieldwork visit begins*, make sure you have (1) decided what your own research focus will be, and (2) thought of ways you can **adapt** or **modify** part of the work that your class will be carrying out ■ Failure to take these steps will reduce the marks you can be awarded	■ You may decide to produce an independent investigation that is entirely separate from the whole class 4-day fieldtrip(s). You will, however, draw on the skills you learn from your fieldtrip(s) when carrying out your unique independent investigation afterwards ■ For example, your class might visit a Field Studies Centre for 4 days in April (Year 12) in order to carry out coastal and places fieldwork. Having participated in this, you may decide to carry out your own independent investigation researching infiltration rates in local parks at home during the summer holiday ■ You might be able to adapt sampling or analytical methods from the class fieldtrip	■ It is possible to combine approaches. For example, your class might visit a Field Studies Centre for 4 days in September (Year 13) to carry out fieldwork exploring the cultural diversity of an urban place ('Town A'). The entire class would participate in primary and secondary data collection for Town A ■ You could then carry out a follow-up study at a second location ('Town B') close to where you live. You could refine and repeat the fieldwork carried out in Town A, but this time by yourself in Town B. This could be done during October half-term ■ Your completed study would give a unique comparison of Towns A and B (note that Town A information must be classed as secondary data, however)

Choosing between a question, aim or hypothesis

Another thing to consider at the planning stage is whether your independent investigation's title will take the form of an **aim**, **question** or **hypothesis**. There is no right or wrong choice to make and each approach has its merits. Table 17 shows examples of these three different approaches. Which one appeals most to you?

Table 17 Aim, question or hypothesis?

	Aim	Question	Hypothesis
	A statement of what your geographical enquiry wants to achieve. A well-thought-out enquiry aim will be geographically sound and achievable	Key research question(s) are used to set up the enquiry. One question may serve as the overall enquiry title, while additional questions are used to help subdivide the written report	A hypothesis is a statement, the accuracy of which can be tested objectively using scientific methodology. You may also be familiar with the use of null hypotheses in science lessons
Examples	■ An investigation into the effectiveness of traffic calming measures in Streatham ■ A survey of how diaspora communities use social networks to maintain contact with families	■ How and why do beach profiles vary along the Penarth coastline? ■ To what extent is altitude the main influence on the distribution of corries in Snowdonia?	■ Most shoppers in Brighton now purchase more online than they do in shops ■ Infiltration rates are higher under coniferous forest cover than under deciduous forest cover

Exam tip

Make sure you choose a question that is geography-focused. An English student may be tempted to study how landscapes are represented in classic novels, but you must produce a geography report, not an English one. The same risk holds for biological studies of the carbon cycle. To ensure your investigation is geographical, ask: *What maps might I draw, other than those to establish the context?* If you answer 'none', your planned work may lack sufficiently spatial elements.

A summary of planning stage expectations

As you plan ahead, Figure 50 provides a useful summary of what to expect in terms of those phases of the independent investigation that require you to work alone and the times when you are allowed to collaborate with other students or seek some teacher guidance. In particular, note that:

■ you will decide your specific enquiry question(s) independently, but this decision can be informed by teacher input and guidance on the broad topic frame (*high level of independence*)

■ you need to be involved in key decisions about what fieldwork procedures and data sources will be used in order to demonstrate active engagement with the process (*medium level of independence*)

■ after the fieldwork has been carried out (working either with a group or alone on a 'solo mission'), you must analyse your evidence and draw conclusions independently (*high level of independence*)

■ you are also expected to work alone to critically evaluate the validity of your evidence and the reliability of your methods (*high level of independence*)

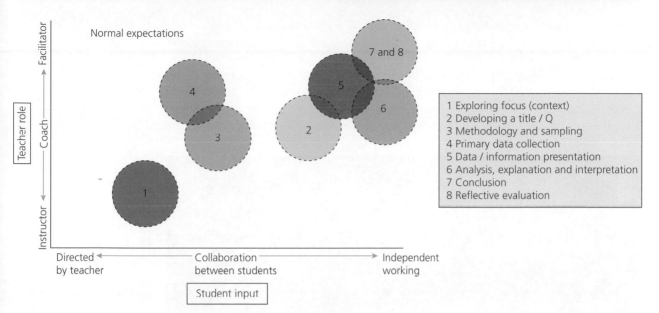

Figure 50 Normal expectations for the independent investigation (source: Nick Lapthorne and Becky Kitchen)

■ Carrying out an independent investigation

The types of data that can be collected, and the range of methods available, are widely covered by many textbooks and websites. In particular, the UK's Royal Geographical Society (RGS), Geographical Association (GA) and Field Studies Council (FSC) have produced quality materials that are, for the most part, freely available and which you can use and adapt.

Collecting data

Independent investigation topics will vary greatly in terms of what kinds of data are collected, and how. For example:

- Many projects will rely mainly on **quantitative data**. This is information that consists of numerical data, for example **secondary data** from the 2011 Census showing inequalities between different electoral wards, or **primary data** collected from a beach showing the range of pebble diameters.

- Some projects will use a greater amount of **qualitative data**. Qualitative methods include semi-structured and open-ended interviews which have been carried out face to face. These interviews are usually recorded with a smartphone, and are later typed up (or 'transcribed') to produce long sections of text which need to be carefully analysed, or 'coded'. Qualitative methods also include the collection and use of images, including photographs, paintings or adverts. These can be analysed or 'deconstructed' using techniques adapted from other academic subjects, including Art, Media Studies and English Literature. Qualitative methods are most likely to be used by students who are carrying out an independent investigation on the topic of Changing places.

■ **Technology** plays an important role in data collection and analysis (Figure 51). Increasingly, geography students make greater use of online databases, **geographical information systems** (**GIS**) and smartphone apps when carrying out research. Again, several detailed resources have been written for A-level geography students by the RGS and GA. These provide useful pointers on how to employ cutting-edge technology in your work. Because technology is advancing rapidly, especially in terms of the way smartphones can support geographical research, make a point of looking for the most recent resources offering advice that you can find.

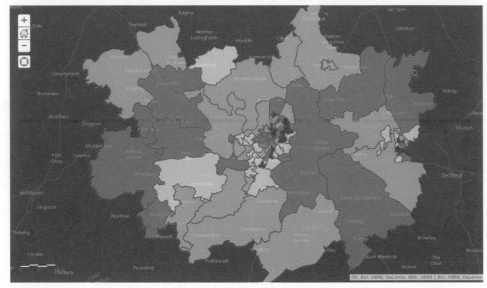

Index of Multiple Deprivation 2010 for West Midlands

- >23,873 to 30,802
- >17,184 to 23,873
- >10,257 to 17,184
- >2,077 to 10,257

Figure 51 The 2010 Index of Multiple Deprivation (IMD) is a useful online data source for place studies (so, too, is the WIMD, or Welsh Index of Multiple Deprivation)

■ It is hard to envisage an independent investigation that does not make use of a **sampling strategy**. Getting this right is one of the most important methodological considerations. Figure 52 shows three different sampling frames for a spatial enquiry. You may also be familiar with particular sampling strategies, such as **random** sampling, **stratified** sampling, **systematic** sampling and **snowball** sampling. It is particularly important to provide a justification of the sample size you decide to work with. Usually there is a sensible balance to be struck between collecting sufficient data to make meaningful generalisations, and not attempting to collect such a large sample that it becomes impossible to complete the work. There are opportunities in the independent investigation for you to reflect on this and other important decisions. *The greater the effort you put into providing a rationale for all of your decisions, the more likely it is that you will score highly according to the marking criteria.*

Fieldwork techniques

The sheer number of possible topics and approaches in A-level geography research and fieldwork means that only a small selection of possible techniques can be covered here. Table 18 will help you begin to focus on the kinds of techniques that you might use once you have identified (1) your broad topic frame and (2) a more narrowly focused research question or hypothesis.

(a) point sampling

(b) line sampling

(c) area sampling

Key

☐ = total population study area

Figure 52 Different sampling frames for a spatial enquiry

Table 18 An overview of possible primary fieldwork techniques and foci

Surveys	■ Many human geography projects make use of surveys. A well-designed questionnaire can generate large volumes of data which can be manipulated graphically and subjected to statistical testing. The focus may be objective measures, such as the distance people travel to visit tourist attractions; or people's behaviour and subjective views about something, such as the desirability of a new urban development
	■ Surveys can also be used in physical geography, for instance when finding out how people's perceptions of local flood risk vary based on their length of residence and personal experience of extreme water cycle flows in their locality
Interviews	■ Some students carrying out an independent investigation in human geography may decide to interview a small number of people at greater length. Semi-structured or unstructured interviews can take the form of a conversation lasting for perhaps 15 minutes. You need to type up and analyse a large amount of text if you adopt this approach. The interview should only be recorded with the consent of the interviewee, and not secretly. This is an important issue of ethics
	■ Not all topics lend themselves well to in-depth interviews. However, a study of the experiences of economic migrants or refugees might be well suited to a small sample of in-depth interviews
Store size measurements	■ Analysis of the size of a beach store and/or its typical sediment size can be carried out in a rigorous way using tape measures, calipers and other equipment that is easy to handle. Some techniques may involve teamwork, such as measurement of the gradient of the land using a clinometer, ranging poles and tape measure (Figure 53)
	■ Carbon stocks in woodland can be calculated using standard equations to estimate the living biomass of trees from the diameter measured at 1.3 m height (see http://tinyurl.com/q5mgxru for an account of this)
Flows and movements	■ Different flows and movements can be studied within physical systems. Infiltration rates can be calculated using a stopwatch to time how long it takes a particular volume of water to drain from an infiltration ring or plastic pipe into the soil. The speed of water in a river channel can be measured using specialist equipment or simple surface floats
	■ Sediment movements over time can be studied using tracers or coloured pebbles, the changing location of which can be mapped over a sequence of days (following their movement by longshore drift)
	■ Precipitation flows can be studied over a period of time using simple rain gauges
	■ Human flows — such as traffic or pedestrians — can be counted and rates calculated
Representations	■ A wide range of visual data can be collected, such as photographs of urban graffiti, or the comments that people have posted about a place on a website such as TripAdvisor
	■ You might even collect examples of old folk songs and music which have been written about a place, or use old newspapers to study how a football team and its fans have been portrayed in the past
Patterns and distributions	■ Patterns and distributions can be measured by sampling data at numerous different sites and recording the information on maps. You can do this electronically by recording data using a smartphone (e.g. to take decibel readings in an urban area). The information can be automatically uploaded to a GIS base map
	■ Large amounts of data can be collected using an environmental quality index (EQI) assessment; each variable recorded can be mapped to show the pattern or distribution

Figure 53 In order to measure slope gradients and profiles accurately, it may be essential for students to collaborate at the data collection stage

■ Completing an independent investigation

Data analysis techniques

Alongside many traditional ways of presenting and analysing data — including hand-drawn maps and charts — you can use GIS and Microsoft Office programs to display data in varied ways, often to great effect.

- Be careful, however, not to overuse flashy and eye-catching techniques which may at first seem to have a 'wow' factor but are in fact less effective at communicating what you want to say than some simpler old-fashioned presentation methods.
- Again, the GA and RGS provide ample support for students on how to use presentation and analysis techniques effectively.
- Table 19 provides a quick overview — as a starting point — to get you thinking about possible ways of manipulating your data once you have collected them.

It is a good idea to make use of statistical tests such as chi-square and Spearman rank correlation, provided it is appropriate to do so and your sample sizes are large enough to justify this (you would not carry out a Spearman rank correlation test for just four or five pairs of data, for example). It is one thing to be able to complete the statistical calculation; it is another to provide a meaningful account of what the test result shows. Make sure you can communicate the findings of any statistical tests in a clear way. When you are discussing **statistical significance**, take sufficient time to provide

The **statistical significance** of a finding means the probability that the results or outcomes are not merely due to chance. Statistical significance of 95% means there is only a one-in-twenty likelihood that the result occurred by chance.

a lucid and easy-to-understand account for the reader. This is important because you may be investigating 'the extent to which' a relationship is true or a scheme has been successful. In which case, the statistical significance level of any testing you carry out could play a pivotal role in determining the quality of your final conclusion and evaluation.

Table 19 An overview of graphical and statistical skills you might consider using

Technique	Tips for using these techniques
Mapping	■ Follow all the important conventions, such as including a scale and north arrow ■ Be careful not to use too many colour classes or to over-complicate a map to the point where any patterns or distributions become more difficult, rather than easier, to see
GIS	■ GIS can help you analyse your data in innovative ways. Unlike paper mapping, GIS contain a database which you can continue to access at any point during your data analysis ■ You can use filters to highlight locations that meet particular criteria (such as urban fieldwork locations where a pedestrian count higher than 50 was recorded) ■ It is well worth taking the time to find out more about this, for instance at: www.ordnancesurvey.co.uk/support/understanding-gis/
Scatter graphs	■ Be wary of drawing a straight best-fit line through a scatter graph. For instance, the relationship between infiltration and precipitation is non-linear, and you should *not* attempt to draw a straight line on a scatter graph showing precipitation and infiltration for rainstorms in a catchment
Charts	■ Try not to over-use overly simple bar and pie charts — after all, these are techniques which even primary schoolchildren can make use of. Instead, think about better ways of *combining* data sets visually, for instance by using grouped or divided bars ■ Rather than showing two separate 'before and after' pie charts, it may be preferable to use two semi-circles drawn back to back ■ Read some quality newspapers and magazines such as the *Financial Times* and The *Economist*, and look at the way their data are presented. There may be some interesting approaches that you can modify and use in your own work
Annotated photographs	■ Do not fill up space in your Independent investigation with large numbers of photographs that lack any annotation or text. It is your job to add value to visual images ■ Overlaying captions and text boxes, labels and arrows on photographs is relatively easy using a program such as Microsoft Word
Less well-used graphical techniques	■ Your WJEC and Eduqas specifications contain a long list of different cartographic and graphical techniques on pages 54 and 49, respectively, in addition to possible statistical calculation techniques. Familiarise yourself with these, and search online for any you are unfamiliar with (an image search will be useful if you want to see examples of a Lorenz curve or logarithmic graph, for instance) ■ This may provide you with inspiration for interesting and less widely used ways to present some of your own data, and can help make your independent investigation stand out from the crowd
Qualitative data	■ Qualitative data can be a challenge when it comes to data presentation. Many academics publish work using qualitative data which read more or less like newspaper articles (i.e. long essays containing quotations from people who have been interviewed). In order to meet the marking criteria for data presentation you should probably attempt to visualise any quotations from interviews. For instance you could overlay speech bubbles on an aerial photograph of the place you have been studying
Spearman and chi-square tests	■ For both of these tests, be careful to perform the calculation accurately and to communicate your findings effectively ■ You may decide to include an extract from a critical levels table in your independent investigation, possibly as an appendix

Accessing the highest levels of the mark scheme

Table 20 shows the criteria that teachers use to award the highest marks for each of the six categories in the assessment grid when marking completed independent investigations.

As you can see, there are frequent references to literature, theories, 'sophisticated' work, and 'effective' evaluation. If you hope to meet these targets in order to access the highest mark bands, then you need to think carefully *from the outset* about the opportunities for sophisticated and reflective geographical thinking that your independent investigation title will provide. For this reason, students are sometimes advised to make use of the **specialised concepts** for geography, such as risk, resilience, threshold and equilibrium. These were introduced on p. 8 and feature frequently in the material covering the water and carbon cycles in this study guide. You might even consider including one of these concepts in the title of your investigation.

- For example, rather than asking the simple research question 'How has flooding impacted on Bewdley?' you might instead ask: 'To what extent is Bewdley a flood-resilient place?'
- This second title definitely opens up more interesting avenues for writing a thoughtful conclusion and evaluation (for which, as Table 20 shows, many marks are potentially available).
- Because resilience is, in any case, a difficult concept to pin down, its use as part of the study title could provide an opportunity to discuss more critically the validity and/or **reliability** of the investigation's results.

Reliability refers to the consistency or reproducibility of a measurement. A measurement is said to have high reliability if it produces consistent results under consistent conditions. True reliability cannot be calculated — it can only be estimated based on knowledge and understanding of the topic.

Table 20 The top bands for the WJEC and Eduqas mark schemes

Context	Methods	Data presentation	Analysis and interpretation	Conclusions and presentation	Evaluation
9–10 marks	13–15 marks	9–10 marks	13–15 marks	9–10 marks	17–20 marks
■ Wide ranging and thorough use of literature sources with a confident theoretical and/or contextual background leading to a well-defined research question ■ Confident and informed understanding of risk/ethical issues	■ Strong evidence of wide ranging and good quality data collection approaches (quantitative, qualitative) relevant to the topic, linked to a well-defined individual research question ■ Practical approaches taken in the field are accurately and well explained and justified ■ Sampling strategy is well designed, explained and justified	■ Wide ranging and accurate use of appropriate qualitative and/or quantitative data presentation methods/ techniques ■ Well selected, applied and wholly appropriate cartographic and graphical techniques to support the analysis of findings	■ Sophisticated analysis and interpretation of findings, clearly showing why they were appropriate and relevant to the research question ■ Demonstrates some individuality and/or insights into links between the study and other aspects of geography	■ Sophisticated and confident summary, drawing convincing and thorough individual conclusions that address the research question and substantiate the analysis and interpretation ■ A well-structured, concise and logical report; accurately references secondary information ■ Spelling, punctuation and grammar used with a high degree of accuracy	■ Highly effective evaluation of knowledge and understanding gained from field observation ■ Perceptive evaluation of each stage of the fieldwork investigation, including the ethical dimensions of the research ■ Perceptive reflections for further research and extension of geographical understanding ■ Considered improvements suggested

The basics of good presentation and submission rules

Finally, make sure you are familiar with all of the rules and instructions that need to be followed as part of the submission process. For example, your specification requires you to produce a word-processed report in Arial, Calibri or Times New Roman (11 point) font, with line spacing set at 1.5. All pages must be numbered. There are other important rules and procedures to follow too: read your specification carefully.

The following provides you with a useful final checklist to follow; several of these 'good practice' tips, as you can see, deal with all-important presentation issues.

Don't forget to…

- choose a tightly focused fieldwork question and, if relevant, a strictly limited number of hypotheses
- link the fieldwork question clearly with your geography specification
- ensure that your proposed work has a clear spatial component, and involves collecting data that you can represent on maps, ideally. Be careful not to carry out what is essentially a biology or economics investigation!
- make sure you know about the 'specialised concepts' — such as risk and threshold — and how they can be used (ask your teacher if unsure)
- personalise any downloaded maps to show the location, choice of topic and/or sample points, following standard geographic conventions such as including a scale and north arrow
- justify (in detail) all the methods used and explain the sampling method(s) employed
- ensure that ample quantitative data are collected for graphs. Limit the application of statistical tests such as Spearman rank correlation to situations where sufficient data have been collected
- be aware of the wide range of graphical techniques and simple statistical tools that are available for your data analysis. Fieldwork investigations benefit when a variety of techniques are used
- avoid including too many extensive tables in your investigation, especially in the sections for methods and evaluation
- incorporate a wide variety of appropriate graphical and mapping techniques in your analysis
- in the analysis, focus on interpreting (not just describing) results and explaining your findings, focusing on any spatial patterns or trends identified
- number and place all the illustrations appropriately in the text, and then refer to them throughout the written analysis
- pay close attention to the assessment criteria and follow the recommended report structure
- print and present your finished report in a light and user-friendly format, following WJEC/Eduqas rules

Further reading

- The Royal Geographical Society (RGS) produces a comprehensive guide to fieldwork, and gives information and instructions for a range of techniques that can be used to carry out geographical fieldwork investigations in different locations and settings: https://tinyurl.com/jcyetxc
- The Geographic Association (GA) provides advice to students at: www.geography.org.uk/resources/studentinformation/
- The Barcelona Field Studies Centre has produced a series of geography resources which are available free: http://geographyfieldwork.com
- The Field Studies Council (FSC) has a website dedicated to A-level fieldwork. This is constantly being updated to match the changing requirements of A-level: www.geography-fieldwork.org
- *Geography Review* published by Hodder Education has a regular feature on geographical skills.
- The Ordnance Survey has a large range of resources based on maps. They range from simple map skills to complex use of GIS: www.ordnancesurvey.co.uk/mapzone/gis-zone

Questions & Answers

About this section

The questions below are typical of the style and structure that you can expect to see in the actual exam papers. Each question is followed by comments, indicated by ⓔ, which list the points that a good answer should include and provide a mark scheme. Student responses are then given, with further comments (ⓔ) indicating how the answer could be improved and the number of marks that would be awarded. (Note that most of the student answers given here would achieve a B grade or A grade overall.) The numbered references included in the student answers indicate the points that specific comments refer to.

Four sample questions are provided. These relate to Section A of the WJEC and Eduqas examinations on Global systems and global governance.

- Questions 1 and 2 are accompanied by a figure or table, which you should use to help answer the data-stimulus questions that follow. Each question is worth 7–10 marks, depending on whether you are a WJEC or Eduqas student.
- Questions 3 and 4 are worth more marks, and are more open, essay-style questions that do not make reference to a figure. These extended response questions are worth 18 or 20 marks, again depending on whether you are a WJEC or Eduqas student.

When the examiners read your work they will have a small grid telling them the maximum marks for each assessment objective (AO) (see p. 6). In the example questions that follow, this grid has been placed after each question part. The official mark scheme will include indications of the content, marking guidance and marks bands. One cannot say what grade any of these answers will obtain because, at the time of writing, boundaries have not been decided.

Sample questions

Question 1

This question follows a format used by both WJEC and Eduqas A-level short structured questions. You have around 12–13 minutes to study Figure 1 and answer both parts of the question (which relates to the water cycle).

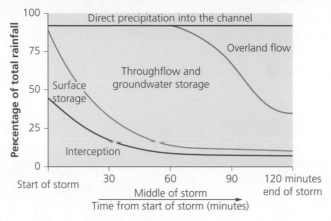

Figure 1 Changes in hydrological processes during a storm

(a) Use Figure 1 to analyse the changing importance of different hydrological processes during a storm.

(5 marks)

(e) Answers might include the following AO3 (skills) points:

■ At the start of the storm, almost all rainfall enters surface storage or is intercepted.

■ Surface storage and interception decline in importance rapidly with time. After the first 30 minutes, their combined importance has declined from around 90% to 30%.

■ Overland flow does not begin until 60 minutes into the storm, and accounts for around two-thirds of total rainfall 120 minutes after the start of the storm.

■ Surface storage and interception account for only 10% of rainfall by the end of the storm.

■ Direct precipitation does not vary in importance (always about 10%).

Level mark scheme

Band	Marks	Use Figure 1 to analyse the changing importance of different hydrological processes during a storm
3	4–5	Well-developed analysis of the changes shown in hydrological processes over time Wide use of evidence to identify changes in the relative importance of the processes
2	2–3	Partial analysis of the changes shown in hydrological processes over time Partial or no use of the resource to identify changes in the relative importance of the processes
1	1	Limited statements with no use of evidence
	0	Response not creditworthy or not attempted

Student answer

At the start of the storm, 10% of rainfall falls directly into the channel, 5% becomes throughflow, 45% is intercepted and 40% enters surface storage. By 1 hour into the storm there have been big changes. Interception has fallen to around 10%, surface storage is now only around 5%, but throughflow (and groundwater storage) has risen to around 75%. After that, overland flow becomes important, and by the end of the storm around 60% of all new rainfall runs off the land as overland flow.

ⓔ **4 marks awarded** This answer describes most of the important changes in a clear way and uses data in a precise and accurate manner, thereby demonstrating skill at handling quantitative data. One minor criticism may be that the answer becomes very 'list-like' in places. Overall, this answer achieves Band 3, but only just. Could you re-write this answer in a way that conveys more clearly the relative importance of the different processes at different time periods? Might it be possible to comment on the speed with which some of the changes in relative importance take place?

(b) Suggest reasons why these changes occur.   (5 marks)

ⓔ Answers might include the following AO2 (applied knowledge) points:

- At the start of the storm event, the vegetation foliage has plenty of capacity to store water. Water can also collect in small hollows and depressions.
- After a few minutes, water begins to be transmitted from the surface to the soil by infiltration. This in turn generates throughflow.
- By 30 minutes into the storm, sufficient rain has fallen onto leaves to exceed the storage capacity of these small 'reservoirs'. Little further rainfall can be stored in this way; stemflow and drip begin transferring new rainfall to the ground quickly.
- By 60 minutes into the storm, the infiltration capacity of the soil begins to decline because of the number of soil pore spaces that are now occupied by water. As the soil becomes saturated, overland flow is generated.
- In this example, the overland flow mechanism is saturation-excess (rather than infiltration-excess).
- The amount of rainfall onto the channel is unchanging because the channel area does not vary.

Level mark scheme

Band	Marks	Suggest reasons why these changes occur
3	4–5	Well-developed suggestions of why the main changes occur Wide use of the candidate's own knowledge to support these suggestions
2	2–3	Partial suggestions of why some changes occur Partial use of the candidate's own knowledge to support these suggestions
1	1	Limited statements with no use of evidence
	0	Response not creditworthy or not attempted

Student answer

At the start of the storm we can assume that the vegetation surfaces are dry and have plenty of capacity to store new inputs of rainwater, especially if the vegetation consists of broad-leaved trees. This is why there is no overland flow and virtually no throughflow to begin with. However, soon after the rain begins, the capacity of leaf surfaces to store further rainfall is exhausted. Also, water which has collected in ground surface depressions begins to soak into the soil and generate throughflow. This is why throughflow becomes the dominant process after 30 minutes. Later on in the storm event, saturation of the soil results in a rapid increase in overland flow because most further rainfall cannot be stored anywhere.

🅔 **5 marks awarded** This is a thorough account that acknowledges the main changes explicitly in addition to the changing relative importance of different processes over time. Although direct precipitation is not mentioned, all of the other main processes are dealt with in a clear and well explained way which also pays close attention to details (the comment about leaf type is good). Overall, this would achieve Band 3.

Question 2

This question follows a format used by both WJEC and Eduqas A-level short structured questions. You have around 12–13 minutes to study Table 1 and answer both parts of the question (which relates to the carbon cycle).

Table 1 Variations in carbon storage (amount and location) for selected biomes

Natural biomes	Total carbon stored globally (GtC)	Where most carbon is stored in the biome
Temperate grasslands	184	Below ground
Tropical rainforests	548	Above ground
Deserts	178	Below ground
Tundra	155	Below ground

(a) Suggest reasons for the variations in carbon storage shown in Table 1. (5 marks)

🅔 Note the command word: this answer asks for suggested reasons (explanations) to be provided using applied knowledge (AO2). A skills-based (AO3) description of the data is *not* required. Suggested reasons might include the following explanatory points:

- Tropical rainforest is by far the largest store (around three times higher than the others) on account of its year-long growing season and constant supply of precipitation.
- Tundra has the lowest storage (155 GtC) because of the lack of light in winter, low temperatures and relatively low precipitation.
- Arid deserts and grassland experience low rainfall as a limiting factor.
- Only tropical rainforests have a climate that can support trees, hence the large above-ground biomass.

- 80% of the biomass of grasses is found in their roots, therefore the largest store is below ground.
- Larger amounts of carbon are stored in the form of methane in a frozen tundra permafrost.

Level mark scheme

Band	Marks	Suggest reasons for the variations in carbon storage shown in Table 1
3	4–5	Well-developed suggestions dealing with variations in biomass carbon storage Wide engagement with the resource and what it shows
2	2–3	Partial suggestions dealing with variations in biomass carbon storage Partial engagement with the resource and what it shows
1	1	Limited statements with no reference to resource
	0	Response not creditworthy or not attempted

Student answer

There is great variation shown here, with tropical rainforests storing the greatest amount of carbon (548 GtC). Tropical rainforests are the only biome where the carbon is stored mainly above ground. All of this reflects the optimum conditions for vegetation growth found between the tropics where the rainforests grow. The other biomes shown all store much less carbon, and carbon is stored mainly in the soil. This reflects low rainfall in all three cases. Temperate grassland receives around 500 mm of rain and the tundra often has less than this. Arid deserts receive less than 250 mm of rain annually.

ⓔ **3 marks awarded** This is a competent attempt to suggest reasons for the main variations shown. The key distinction between the tropical rainforest and other biomes has been highlighted, although the suggested explanation is not clear (what does 'optimum' mean?). Good use of applied knowledge is shown in the way the student correctly identifies annual rainfall totals as possible limiting factors. It is a pity that no reason is offered for why most carbon is stored in the soil in three cases. Overall this would achieve a higher Band 2 mark. What could you add to improve this answer further (but without writing a great many more words)?

(b) Using examples, explain the difference between positive and negative feedback in systems.

(5 marks)

ⓔ This AO1 question is entirely independent of Table 1. Answers might include the following points:

- Positive feedback involves an acceleration of change in a system on account of amplified changes and knock-on effects.
- Negative feedback involves adjustments within a system that can cancel out or neutralise the disruption that has taken place.
- Examples of positive feedback include the melting of Arctic ice cover because of albedo changes and because of permafrost melting which releases large amounts of methane gas thereby raising atmospheric temperatures further.

- Examples of negative feedback include greater cloud cover resulting from evaporation following snow or ice melting; light is reflected, which may help reduce ground temperatures, thereby allowing ice to reform.
- Negative feedback restores equilibrium within a system, whereas positive feedback can cross a system threshold, thereby taking the system to a new state.

Level mark scheme

Band	Marks	Using examples, explain the difference between positive and negative feedback in systems
3	4–5	Well-developed explanation of how positive and negative feedback differ Sustained use of supporting examples
2	2–3	Partial explanation of positive and/or negative feedback Some use of supporting example(s)
1	1	Limited statements with no use of evidence
	0	Response not creditworthy or not attempted

> **Student answer**
>
> Permafrost melting is a good example of positive feedback leading to the acceleration of climate change. Vast amounts of methane (a gas containing carbon) is stored in permafrost below the Arctic tundra at high latitudes in Russia and North America. As global temperature rises, permafrost is melting, particularly in coastal regions of Alaska. This is releasing methane into the atmosphere, which is a potent greenhouse gas. As a result, even more permafrost will melt. The result is a cycle of change which speeds up dangerously. In contrast, negative feedback is the natural capacity of physical systems to regulate themselves and restore equilibrium. If one process speeds up, this can cause another process to slow down.

ⓔ 4 marks awarded This answer conveys understanding of positive feedback extremely well using a strong supporting example. In contrast, the explanation of negative feedback is not clear and lacks a supporting example (despite some good use of terminology). However, the answer manages to make the difference between the two processes clear and transparent to the reader (one process accelerates change whereas the other neutralises it). On balance, this answer might deserve a low Band 3 score.

Question 3

This essay question follows the format used in WJEC questions. You have around 25 minutes to plan and write your essay (which relates to the carbon cycle). Note that in the WJEC examination, there is choice of two essay questions (maximum mark of 18). One essay will relate to the water cycle only. The other essay will relate to the carbon cycle only, or to links between the water and carbon cycles.

To what extent is human activity the main cause of changes in the size of global carbon stores?

(18 marks)

ℯ Answers might include the following points:

- Human causes of negative changes in the size of carbon stores, e.g. deforestation and agriculture.
- Human causes of increased storage, i.e. afforestation.
- The indirect effect of industrialisation on carbon storage in biomes via anthropogenic carbon emissions and warmer temperatures.
- Natural causes of climate change and their effects on carbon storage in biomes and the oceans.
- Other natural causes of changing carbon storage, e.g. plate tectonic movements/Himalayan uplift.
- The extent to which human activity is the main cause of change, either now or in the past (AO2).
- The extent to which human activity creates positive or negative changes (AO2).
- The extent to which natural feedback processes may amplify changes triggered by human activity (AO2).

Good answers may utilise the specialised geographic concepts of feedback, systems or thresholds.

Level mark scheme

Band	Marks	To what extent is human activity the main cause of changes in the size of global carbon stores?	
		AO1	AO2
3	13–18	Mostly accurate knowledge and understanding of a range of causes	Well-developed evaluation of the extent to which human activity is the main cause
2	7–12	Partial knowledge and understanding of a range of causes	Partial or unbalanced evaluation of the extent to which human activity is the main cause
1	1–6	Limited knowledge and understanding of a range of causes	Limited or unbalanced evaluation of the extent to which human activity is the main cause

Student answer

Globally, carbon can be found in the air, land and water. Each of these domains serves as a carbon store. Carbon dioxide and methane are carbon-based gases which are stored in the atmosphere. Carbon exists in vegetation and soil below it. Carbon is also stored in water in a dissolved form and is present in rocks and sediments under the ocean and on land. Although the amount of carbon in the global cycle remains constant over time, changes take place in the relative amount which is stored in different ways. This essay will discuss the relative importance of human and natural causes of these changes. [1]

Since the Industrial Revolution began in 1750, the amount of carbon stored in the atmosphere has risen from around 280 ppm to its current figure of 410 ppm. This is the result of many different human activities including: deforestation and the burning of biomass; fossil fuel use by cars and industries; and cement manufacturing. Carbon has also been introduced to the atmosphere as a result of increased methane emissions from cattle. In recent years, rising affluence in Asia (linked in part to globalisation) has lifted hundreds of millions of people out of poverty, who now can afford a meat and dairy diet. As a result, far more cattle are being reared. Deforestation in Brazil has also been driven in part by Asia's new wealth. Much of the cleared land is used to grow soya crops which are exported to China. Many different interrelated human activities have therefore played a role in increased atmospheric storage of carbon. [2]

Rising carbon emissions are responsible for the slight rise in global mean surface temperature which has already occurred since around 1900. IPCC projections suggest that unless something is done, global temperature may rise by between 2°C and 5°C by 2100. The actual and projected knock-on effects of this for carbon storage are complex and include changes in the pattern of global biomes and carbon storage. If desertification occurs and large areas of grassland are lost, for example in the Sahel or US Prairies, carbon storage will decrease both above and below ground. On the other hand, the highest temperature rises are expected at high latitudes close to the current tree line, which marks the boundary between boreal forest and the Arctic tundra. If the tree line moves north, then large areas of land may change from tundra to boreal forest. This would result in much greater amounts of carbon being stored in vegetation biomass at these latitudes. However, thawing of the tundra soil would release methane gas which is currently locked away in the permafrost. This would release carbon into the atmosphere. [3]

It is not just human activity that causes changes in carbon storage, however. Earth's climate has changed many times in the past. Geologists believe that very cold Cryogenian ice ages occurred around 650–750 million years ago. Much of the Earth was covered with ice and snow: the cryosphere was much larger than today. Global patterns of vegetation and soil carbon storage would have been vastly different. Large areas of the planet which currently support vegetation would have supported none. The cold temperatures would also have affected marine life and the amount of carbon stored in marine food webs. However, the amount of carbon stored in the ocean water in a dissolved form actually increases when water temperatures are lower, therefore it is possible that more carbon was stored in the oceans during colder climatic periods in the past. [4]

In contrast, the Earth experienced hothouse conditions around 30–60 million years ago. The poles were probably ice-free and sea level would have been much higher. Therefore, ocean storage of carbon might have been higher on account of the greater volume of water, although this might have been offset by a warmer temperatures meaning that the amount of carbon stored per litre of water would have been less. [5]

Another important natural cause of changing carbon storage is plate tectonic movement. Antarctica's arrival at the south pole and the formation of the

Antarctic ice sheet caused global sea level to fall by an estimated 70 m, Atherefore reducing the size of the ocean carbon store. The uplift of the Tibetan Plateau and the Himalayas is another important geological episode that affected carbon storage. These mountains are composed of limestone and were originally sea floor deposits that have been thrust upwards into the atmosphere by the convergence of two tectonic plates. The subsequent erosion of these mountains — and the removal of carbon in solution after carbonation has taken place — has reduced the amount of carbon stored there. [6]

In conclusion, humans were clearly not to blame for the majority of changes in carbon storage over time which pre-dated their arrival on Earth. Instead, it is natural climate change and plate tectonic movements that have driven major episodic changes in carbon storage over hundreds of millions of years. More recently, though, humans have arguably had the greatest impact on changing carbon storage. In fact, many geographers describe the current age we live in as the 'Anthropocene' to reflect this. Industrialisation and globalisation have caused an unprecedented increase in atmospheric carbon storage, with many knock-on effects for ecosystems and oceans. Humans may, however, attempt to mitigate these changes through strategies such as afforestation. Therefore it may be that in the future human activity will continue to bring changes in carbon storage but in a more positive way. [7]

e [1] This introduction is a good length and establishes key concepts and topics for discussion. The final sentence shows that the student intends to present his or her ideas as a discursive essay, which is essential if AO2 marks are to be accessed.

[2] This paragraph covers a wide range of ideas without going into too much depth on any one point, which is a good approach to take. Specific and accurate facts and examples are used, and there is reference to two key concepts: globalisation and the idea of interrelationships in geography.

[3] In this paragraph, the answer continues to be well focused on carbon storage (it could be easy to forget to mention carbon when discussing vegetation changes caused by climate change). There is good use of terminology, such as the IPCC and desertification, and specific examples of biomes are provided. It is especially good to see the complex and uncertain nature of changes in carbon storage being acknowledged (using the example of permafrost melting).

[4] This essay is well structured. The discursive counter-argument begins here, approximately midway through the essay. Again, the ideas are supported by details, and the possible complexity of the changes are briefly discussed.

[5] This brief paragraph is useful because it shows that changes in carbon storage can be both positive and negative.

[6] Tectonic movements are shown to have enormous impact on carbon storage, greatly strengthening the student's counter-argument against the proposition that human activity is the main cause of change.

[7] This conclusion is an appropriate length for a student who is aiming to reach the highest marks. It is essential that the AO2 criterion (evaluation) is met alongside the AO1 criterion (knowledge and understanding). The conclusion goes far beyond a simple 'yes or no' answer and discusses the relative importance of human activity during different geological periods. There is also a brief acknowledgement that humans may in the future undo some of the changes they have already caused.

e **18 marks awarded** Overall, this is an excellent answer. A high level of achievement is reached for both AO1 and AO2. With regard to AO1, a wide range of themes are covered in an informed and well-evidenced way, with a good balance of human and non-human causes included. The answer goes far beyond merely explaining human and non-human causes: a proper discussion has been produced that explicitly evaluates the extent to which human activity is the main cause. The use of a geological timescale is particularly effective in shaping the discussion.

Question 4

This essay question follows the format used in Eduqas essay questions. You have around 20–25 minutes to plan and write your essay (which relates to both the water and carbon cycles). Note that in the Eduqas examination, there is a choice of two essay questions (maximum mark of 20). These essays are designed in a way that requires you to draw *equally* on your knowledge and understanding of *both* the water and carbon cycles. You are expected to write about *both* topics in the single essay you choose.

'Precipitation plays an equally important role in both the water and carbon cycles.' Discuss.

(20 marks)

e Good answers might include the following points:

- The role of precipitation in the water and carbon cycles.
- Precipitation as an input in the global water cycle and local drainage basin systems.
- Precipitation intensity and duration, and the consequent occurrence of throughflow and overland flow.
- Precipitation falling as snow and its subsequent surface storage for periods of time, and the associated concept of a steady-state equilibrium (with system inputs and outputs balancing over time).
- The role of precipitation in transferring carbon from the atmosphere to the land.
- The role of precipitation in chemical weathering (carbonation) of sedimentary rocks, and the subsequent transfer of carbon in solution to rivers and the ocean.
- Precipitation and biome growth, and the link between annual precipitation and the size of carbon stores in tropical rainforest, temperate grassland and other biomes.

- The extent to which precipitation is equally important for both cycles (AO2).
- The importance of different aspects of precipitation, i.e. type, amount, intensity, duration (AO2).
- The importance of precipitation for water and carbon cycles in different local contexts (AO2).

Level mark scheme

| Band | Marks | 'Precipitation plays an equally important role in both the water and carbon cycles.' Discuss. | |
		AO1	AO2
3	14–10	Accurate knowledge and understanding of the importance of precipitation	Well-developed and structured evaluation of its role in the water and carbon cycles
2	8–13	Partial knowledge and understanding of the importance of precipitation	Partial or unbalanced evaluation of its role in the water and carbon cycles
1	1–7	Limited knowledge and understanding of the importance of precipitation	Limited or unbalanced evaluation of its role in the water and carbon cycles

Student answer

Precipitation can fall as rain or snow, and can be the result of the uplift of air due to orographic, frontal or convection mechanisms. Precipitation is an input into the drainage basin water system and can vary in its type, amount, duration and intensity. Precipitation is very important for the water cycle because without it the water cycle could not operate. Water which has flowed from rivers into the sea and evaporated into the atmosphere is later returned to the land by precipitation, thereby completing the cycle. [1]

The type of precipitation that falls plays an important role in determining what kind of water flows take place in a drainage basin. Very high intensity rainfall sometimes occurs in the UK in summer during intense convection thunderstorms. Arid areas also experience high intensity rainfall on the rare occasions when it does rain. High intensity rainfall can exceed the infiltration capacity of the soil. This means that even if the soil is dry, excess rainfall which cannot infiltrate begins to flow rapidly downhill under gravity. This is called overland flow or surface runoff and it can lead to a flashy hydrograph with a high peak discharge. [2]

In contrast, low intensity rainfall can also lead to flooding if it continues for long enough. Over time, infiltration into the soil begins to cause soil saturation (the pore spaces in the soil become filled with water). After many hours of steady but low intensity rainfall, the soil is completely saturated. Any further rainfall cannot infiltrate and overland flow occurs instead. However, there will be a much longer lag time associated with this type of rainfall event compared with high intensity rainfall. [3]

Precipitation may also fall as snow. This can be stored on the ground surface for long periods of time. For instance, much of the precipitation in winter in the Rockies and Winter River mountains of the USA falls as snow. It is only when the snow melts in spring that the water is released and can flow into tributaries of the Colorado River. The type of precipitation therefore plays an important role in the water cycle. [4]

Precipitation also plays an important role in the carbon cycle. Carbon dioxide which occurs naturally in the atmosphere can dissolve in rainwater to form a weak carbonic acid (H_2CO_3). When this weakly acidic rainfall comes into contact with carbonate rocks such as chalk and carboniferous limestone, a chemical reaction takes place which produces calcium bicarbonate, which is soluble. Globally, 0.3 billion tonnes of carbon are transferred from rocks to the oceans every year by chemical weathering. This is an important flow in the carbon cycle at a global scale. [5]

Subsequently, the soluble calcium bicarbonate which reaches the oceans is used by marine organisms to create shells. When they die, the shells sink to the bottom of the ocean where they eventually become part of new sedimentary rocks which are moved as part of a slow carbon cycle towards plate boundaries. There, the melting of the rocks by subduction eventually releases carbon dioxide back into the atmosphere. [6]

Precipitation is vital too for the growth of vegetation, which is another important carbon store. Tropical rainforest contains large amounts of carbon in its biomass. Globally, 550 GtC is stored in tropical rainforest biomass and soil, and large forest trees typically store 180 tonnes of carbon per hectare above ground. Other global biomes store large amounts of carbon too, but could not do so without precipitation occurring. The lowest levels of carbon storage are found in the desert biome where rainfall is lower than 250 mm per year. Heavy rainfall in some ecosystems helps vegetation to grow but can also leach soils, which reduces the amount of carbon that can be stored there. There is relatively little organic carbon stored in the soils below the tropical rainforest because of the heavy rainfall. [7]

In conclusion, precipitation is very important for the water cycle and also for the carbon cycle. But perhaps overall it is most important for the water cycle. [8]

e [1] This introduction indicates already that the student has little understanding of the AO2 requirements of the question. The introduction to a high-scoring response is likely to begin by defining key terms *and also* by indicating what the structure of the discursive argument will be (i.e. 'This essay will discuss the water and carbon cycles in turn and will seek to compare the role of precipitation in both cycles to see if it is equally important or not.'). Some of the content of this paragraph is 'nice to know' rather than 'need to know' information and could have been omitted.

[2] This paragraph shows good recall and understanding of the role of precipitation intensity and scores well according to the AO1 criterion.

[3] This paragraph shows good recall and understanding of the role of precipitation duration. It deserves credit according to the AO1 criterion.

[4] This paragraph shows understanding that precipitation can sometimes serve as a water store, rather than a flow. An opportunity has been missed, however, to reflect on this explicitly (i.e. to discuss the importance of precipitation as *both flow and store* in the water cycle).

[5] The carbon cycle has not been mentioned at all in this essay until the current paragraph, which suggests poor planning and poor understanding of the AO2 criterion. The AO1 recall shown here is very good, however.

[6] This paragraph shows sound understanding of the slow carbon cycle but seems to be losing focus on the role of precipitation.

[7] This is a strong paragraph. Evidence is used well to support a number of good points about the importance of precipitation for biomass and soil carbon storage. On the basis of this paragraph, the candidate would achieve high marks in a purely AO1 test.

[8] This is a weak conclusion to a discursive essay, which emphasises further the student's poor understanding of the assessment criteria used. Overall, the best fit for this answer is a Level 2 mark which factors in strong AO1 performance and weak AO2 performance.

ⓔ 11 marks awarded Overall, this essay scores well at AO1 because a range of relevant scenes are recalled and presented in a well-evidenced way which uses terminology correctly. However, it scores poorly at AO2, suggesting that the student is not familiar with the requirements of the examination. The introduction makes no reference to any argument. The conclusion merely restates the question and asserts — without any justification or evidence — that precipitation is more important for the water cycle. This essay would therefore only receive a mid-Band 2 mark. Can you think of how to improve it? Take a look at the introduction and conclusion to the student response for Question 3 (although this is an examination question for the WJEC course, the assessment objectives are identical). The student who answered Question 3 has a much better understanding of AO2 requirements.

Knowledge check answers

1 Climatic factors affect where cryosphere storage takes place. Ice accumulation takes place at high latitudes and high altitudes where temperatures are low. Equatorial regions with high rainfall develop extensive river drainage systems, including freshwater lakes; in contrast, arid regions where rainfall is lower than 250 mm/year may have little in the way of surface water storage.

2 Variations in flows and stores will of course depend on location. In equatorial regions, flows and stores may be relatively constant throughout the year, whereas in a seasonal climate, such as the Mediterranean climate, there are marked differences. For example, very little rain falls during summer in southern Spain and river valleys are dry for large parts of the year. Groundwater stores become greatly depleted. In contrast, water flows and stores increase in size during the winter months. Seasonal patterns can be observed in the UK, especially southern England, although high levels of rainfall and river flow have occurred during summer months in some recent years.

3 Surface runoff is a fast movement because the water can run unimpeded over the ground. The fastest speeds are reached on steep and unvegetated slopes. Throughflow is a much slower process because water has to flow through a matrix of tiny pore spaces. The smaller the pore spaces, the slower the movement (clay soils have tiny pore spaces and very slow throughflow rates). The exception to this is throughflow via pipeflow — water is able to flow along old animal burrows and root channels at speeds equivalent to overland flow. Groundwater movements can be fast if large joints and cracks are present in the rock: limestone areas sometimes have underground river systems within the rock itself. Water movement through rock pores varies according to the rock type and depth (rocks at greater depth are more compressed and have smaller pores). Finally there is greater variation in the speed of stream flow because of differences in stream size and efficiency.

4 Biomes include tropical rainforest, coniferous evergreen forest, temperate deciduous forest, tropical grassland, temperate grassland, desert, Mediterranean vegetation and Arctic tundra. Interception is greatest where broad-leaved trees form a canopy and there are multiple understoreys too. Interception storage lessens in the winter for deciduous forest, however, in comparison with evergreen tropical forest. Grassland interception varies according to the height of the grass, which is 2 m in tropical regions but under 1 m in temperate regions.

5 Physical factors were more important in the past before widescale settlement took place in the interior of the USA after the arrival of Europeans. The key influences were seasonal variations in precipitation and evaporation; in particular, seasonal patterns of snowfall and snowmelt in the Rocky Mountains contributed to a notable spike in discharge in early summer. Since the 1900s, human factors have dominated the regime: the construction of large dams has evened out the annual pattern of discharge.

6 An account of both mechanisms is provided on pp. 29–30. Make sure you study both carefully and create revision lists and notes for yourself which summarise key points, using specialised terms wherever you can. In particular, pay attention to the different roles played by ice crystals and water droplets in the Bergeron process, and the role of super-sized condensation nuclei in the collision process.

7 Aquifers can develop where the local geology and water cycle flows create favourable conditions for water storage. Percolating rainwater can collect in the water-bearing porous rocks such as chalk, provided an impermeable layer of rock lies below. However, many aquifers have been depleted by human activity. World population has grown from 1 billion to 7 billion since 1900. In recent decades, millions of people have escaped poverty, particularly in Asia. This increases the size of an average person's water footprint, measured in terms of the water used per capita for drinking, cleaning, and also consumption of water 'embedded' in food and products. As a result, many aquifers such as the one below Mexico City have suffered overabstraction.

8 You may have studied subduction at GCSE or as part of the A-level course already. Be sure to refresh your knowledge of the processes that drive subduction. One theory is called convection. The Earth's core temperature is around 6,000°C. This causes magma to rise in the mantle and sink towards the core when it cools. The currents flow beneath the lithosphere, building up pressure and carrying the plates with them. Another accepted explanation is called ridge push and slab pull. The melting of carbonate rocks occurs when a dense ocean plate subducts beneath a continental plate. As it subducts, the plate melts.

9 You may have studied the trophic levels of an ecosystem as part of your GCSE course or in Biology, and it is worth refreshing your knowledge of this topic. Key points to remember are that herbivores are primary consumers that eat plants, and carnivores are secondary consumers that feed on herbivores. However, many parts of plants and animals are not eaten by the animals that feed on them. Carnivores do not eat all of their prey. Also, much of what animals eat is excreted. Finally, energy is lost at each level: hunters use a lot of kinetic energy chasing their prey. These and other factors influence how much carbon is transferred directly between trophic levels.

10 You may have studied this as part of your GCSE course. The influence of latitude on climate is a key

geographical understanding. At very high latitudes, because of the Earth's tilt, there is no sunlight for some months of the year; but at other times of year the sun does not set.

11 If the soil is saturated, this means all pore spaces that are sometimes occupied by air are now occupied by water. In low-relief areas where peatland has formed, water cannot flow laterally and will be retained if there is impermeable geology below. As a result, water will collect on the surface and generate overland flow soon after rainfall begins, especially at times of the year when it has rained previously.

12 This is a chance for you to review your notes for Changing places, assuming you have already started this topic. It has been important for rural regions of the UK to seek to diversify their economies in recent decades on account of rising unemployment in a mechanised agricultural environment. Speciality food and drink has been seized upon as a possible basis for new startup businesses. In some parts of the UK, where peat has been burned traditionally, local food and drink industries — such as salmon or whisky production — choose to make use of peat smoke as a traditional 'heritage' flavour. Large supermarkets such as Tesco and Waitrose source peat-smoked products from these local rural areas, helping to build links between different places, and between producers and consumers of food.

13 Carbon dioxide emissions are associated with forestry removal, fossil fuel burning and cement production. Methane emissions are associated with cattle ranching and other forms of agriculture.

14 Your approach to this Knowledge check task will depend on how you have been taught these topics. Numerous suggestions have been made in the text which can be discussed with your human geography and physical geography teachers.

Self-study task answers

1 Temperature remains relatively constant throughout the year between 27°C and 30°C, with a slightly warmer season between March and May. In contrast, precipitation shows an uneven pattern. In total, around 3,000 mm of rain falls each year. Incredibly, 2,000 mm of this falls in June and July. Virtually no rain falls between November and April each year.

2 The precipitation for river A is much lower than for river B. In total, river A receives less than 1,000 mm of precipitation annually, with slightly lower levels between February and June. Higher precipitation occurs in the autumn and winter. River flow corresponds poorly with the rainfall pattern, with the maximum river flow level reached in April. This suggests that winter rainfall has taken several months to be transmitted through the basin. This is suggestive of an extremely porous geology such as chalk, which transmits water towards rivers via the slow process of groundwater flow. In contrast, river B has perhaps 1,500 mm of precipitation annually, with a notable winter peak. It is likely that this basin lies in a northern or western region where precipitation is generally higher. River flow corresponds well with precipitation, suggesting a rapid response indicative of impermeable geology which promotes rapid overland flow and relatively rapid throughflow. There is evidence of snow falling and melting in March and April.

3 The infiltration rates shown vary from zero (in clay) to 57 mm per hour (for well-established pasture). All other land uses have a lower infiltration rate of 13 mm per hour or less. There is greater variability in the infiltration rates for different soil types, particularly sand. Infiltration varies according to land use on account of the vegetation cover which is present (which may promote interception) and the degree of compaction of the soil (and the number of pores which water can soak into).

4 Basin A is narrow and elongated; water from each tributary arrives at the measuring point in a sequential manner. Therefore a steady discharge results. In contrast, the radial shape of basin B ensures that water from all of the contributing tributaries converges at the measuring point simultaneously. The result is a steep rise in discharge.

5 Storm A generates the highest peak discharge despite its relatively low amount of rainfall. This is explained by the high maximum intensity of 10 mm per hour. Most likely, this generated infiltration-excess overland flow resulting in a very short lag time. Storm B was the smallest storm in terms of total rainfall. However, preceding discharge was relatively high, suggesting the soil may have been partly saturated. As a result, some throughflow or overland flow may have been generated, and so peak discharge was not the very lowest of the four storms. Storm C was the largest in terms of total amount. However it lasted for a long time — 10 hours — and occurred at a time of year when the preceding discharge was extremely low, i.e. most likely summer. Therefore we can assume that most of the rainfall was able to infiltrate dry soil during the first hours of the storm. Only much later did the soil become saturated, resulting in saturation-excess overland flow and a high peak discharge. Storm D lasted for 16 hours. Despite having the highest preceding discharge, it resulted in the lowest peak discharge. Most likely, the extremely low intensity allowed all of the rainwater to infiltrate the soil or evaporate from vegetation surfaces.

6 In an unurbanised catchment, the magnitude of flood events does not exceed 4 cubic metres per second and these high discharge rates only occur rarely, e.g. once every 10 years. In contrast, a basin which is

100% sewered and 60% impervious (i.e. covered in cement and tarmac) generates floods of much higher magnitude, exceeding 9 cubic metres per second or more. Moreover, these high discharge rates occur every 1–2 years.

7 In November, around 90 mm of precipitation falls as snow. In June, around 125 mm of rainfall occurs. In November, outputs equal just 40 mm of precipitation equivalent (as runoff and evapotranspiration), whereas in June outputs of more than 275 mm are recorded. Clearly there is an imbalance occurring in both months which does not suggest equilibrium. However, if a longer-term view is taken of the drainage basin over the year as a whole, then a state of equilibrium has been reached. The excessive net output in June compensates for the high net inputs recorded during the winter months.

8 The Aral Sea has decreased enormously in size. The large body of water in the centre of the image for the year 2000 has vanished in 2015. All that remains are small isolated lakes flanking the west and north of an area which presumably was once completely covered with water.

9 Transfers can be achieved in seconds by the diffusion of carbon dioxide (it dissolves in the surface of the water). Carbon dioxide is also absorbed by phytoplankton through photosynthesis every second of the day. Transfers between phytoplankton and other organisms, such as larger fish, may take place on longer timescales (as different species hunt and feed on each other). Carbon is also present in the ocean water in the form of soluble carbonate which is used by marine organisms to create shells. Coral reefs grow over decades and centuries in this way. Shells which are deposited as sediment on the ocean floor become lithified as part of a slow carbon cycle that takes place over millions of years.

10 Use this table to help you complete this task. A few words and phrases have been added to get you started.

	Pathways	Processes
Shorter timescale	Atmosphere to vegetation Vegetation to atmosphere	Photosynthesis
Longer timescale		
Local (spatial) scale	Vegetation to atmosphere	Forest fire
Continental or global (spatial) scale	Land to oceans	Chemical weathering

11 Tropical rainforest is found in a band circling the equator between the Tropics of Cancer and Capricorn. It is present in South America, West Africa and Indonesia. There are also patches of tropical rainforest in northern Australia, Madagascar, and the coast of southeast Asia. In comparison, temperate grassland is found north and south of the areas where tropical rainforest is found. It is also found in the interiors of continents — compared with the coastal distribution of rainforest in some parts of the world. Temperate grassland is located at latitudes of around 30–50°N, with some small isolated pockets around 30–40°S.

12 The highest levels of storage are found in tropical forests (547.5 GtC). Other forest biomes have relatively high values in excess of 300 GtC, and tropical grassland comes close to this (285 GtC). Deserts, temperate grassland and tundra all have relatively low values between approximately 155 GtC and 184 GtC.

13 The biome store has reduced greatly in size from 18 to 43 tonnes of carbon per hectare. There has also been a marked reduction in the soil storage, from 64 to 12 tonnes of carbon per hectare stored in below-ground biomass. Major changes can be observed in the relative importance of carbon flows too. Absorption by photosynthesis has more than halved. However, carbon outputs have increased greatly. Respiration is far less important and only one-quarter what it used to be. Burning, 'decay of slash' and soil erosion now contribute to a new emission of 25.1 tonnes of carbon per hectare per year.

14 This is a research task.

15 Carbon flux varies from a net loss of 22.4 tonnes of carbon dioxide per hectare per year in cultivated grass to a net gain of 4.1 tonnes of carbon dioxide per hectare per year in undamaged peatland. The majority of land uses shown have a small loss: overall, six of the seven land uses shown display a loss. This suggests that any interference with peatland in its natural state harms its capacity to store carbon. As soon as drainage is improved, plant material begins to decompose, which releases carbon dioxide. Some of the land uses shown may also result in erosion of the peat by overland flow and gullying.

16 Precipitation saturates the leaf litter and can speed up the rate of decomposition. The impact of rain droplets may be sufficient to help material fragment. By keeping the soil below the litter moist, precipitation also encourages earthworm activity. Overland flow could play an important role, physically washing large amounts of litter or decomposing organic matter into streams or sewers in urban areas. Humus that remains on site in the soil can store water. Humus can hold the equivalent of 80–90% of its weight in moisture, and therefore increases the soil's capacity to withstand drought conditions.

17 Your approach to this task will depend on how you have been taught these topics. Numerous suggestions have been made in the text which can be discussed with your human geography and physical geography teachers.

18 The highest scenario suggests temperature may rise by an estimated 4°C (this appears to be the best estimate from a range of 3–5°C). The lowest scenario suggests a rise of 1.5°C (the best estimate from a range of 1–2°C). The uncertainty which is shown about what will happen in the future is based partly in our imperfect understanding of how physical systems will respond to a rise in GMST.

As you will by now understand, positive feedback effects are complex, and there is uncertainty over the degree to which oceans and ecosystems will serve as carbon sinks for future emissions. There is also uncertainty over the degree to which the world will continue to industrialise, and the extent to which globalisation will continue to lift people out of poverty, or not. Therefore estimates vary widely on what humanity's energy needs will be in the future. We also do not know whether fossil fuels will always be used to meet increased demand for energy, or whether technological developments will provide cheap and reliable renewable energy on a large enough scale.

Index

Note: page numbers in **bold** indicate defined terms.

A

ablation   **12**
accumulation   **12**
acidification of oceans   61, 68
adaptation   8
afforestation   51
agricultural drought   34
agriculture   36, 51–52
air uplift   28, 29
albedo   **66**
anaerobic conditions   **52**, 54
Antarctica   13, 14, 61
antecedent conditions   **25**
aquifers   **35**, 36, 37–38
Aral Sea   36–37
area sampling   78
aridity   **14**
artesian aquifers   37–38
assessment objectives (AOs)   6
atmosphere   11, 58–59

B

bankfull discharge   24
baseflow   22, 24
Bergeron-Findeisen process   29–30
biological carbon pump   42, 43
biomass   **41**
biomes   **20**
    carbon flows and stores   46
    carbon stores   45–52
    ecosystem carbon storage   45–49
    global biome map   47
    human activity changes   49–52
biosphere   11
blanket peatlands   54
blue water   11
boreal (coniferous) forest   45
bunch grasses   49

C

capillary water   20
carbonation   **43**
carbon cycle   **39**
    atmospheric carbon storage   58–59
    carbon emissions and energy
        budget   59–60
    desertification   62
    feedback   63–71
    global carbon cycle   39–44
    inputs, outputs, stores and
        flows   39–40
    links with water cycle   58–63
    pathways and processes   40–44
    sample questions   94–97
    slow carbon cycle   43–44
carbon cycle pump   **41**, 42
carbon market   **56**
carbon offsetting   51
carbon sequestration   40, 42, 51,
    55, 62
carbon sinks   **56**
carbon storage
    atmospheric carbon storage   58–59
    different biomes   45–52
    ecosystem carbon storage   45–49
    global carbon cycle   39, 40
    human activity changes   49–52
    impacts on water cycle   60–61
    peatlands   52–57
    sample questions   88–94
    temperate grasslands   48–49
catchment hydrology   15–21
cation exchange   45
causality   8
channel discharge   21
channel store   19
charts   81
chi-square tests   80, 81
climate change   14, 35, 58, 60, 62,
    64, 68–70
closed system   10
collision process   29, 30
Colorado River   22–23
condensation   10, **27**, 28
condensation nuclei   27
convectional rainfall   28, 29, 61
crop management   52
cryosphere   **10**, 11, 12, 66–67, 69

D

dams   22, 23
decomposition   41
deforestation   32, 49–50, 62
desertification   **62**, 65
distributions   79
drainage basins   15–21, 23, 26–27,
    33–34
drainage density   25, 26
drought   33–35
Dust Bowl   51

E

ecosystem carbon storage   45–49
ecosystem services   **56**
Eduqas examination   5–6
El Niño southern oscillation
    (ENSO)   **14**
energy budget   **59**, 60
equilibrium   8, 63
eustatic sea level change   **61**
evaporation   21, 25, 61
evapotranspiration   21, 33
excess runoff   28, 30–32
extreme weather events   **60**

F

falling limb   24
feedback   8, **63**
    carbon and water cycles   63–71
    cryosphere feedback   66–67
    equilibrium and thresholds   63–65
    impacts   65–68
    implications   68–70
    methane feedback   67
    terrestrial and marine carbon
        feedback   67–68
feeder–seeder mechanism   **29**
fen peatlands   54
field capacity   20, 21
Field Studies Council (FSC)
    77, 84
fieldwork
    carrying out   77–79
    completing   80–84
    planning   71–77

techniques   78–79
flashy hydrographs   25, 27
flooding   15, 18, 27, 30, 32, 60
flows   **10**
  carbon cycle   39, 46
  drainage basins   16, 17–18
  feedback   68
  fieldwork techniques   79
  water cycle   10–11
forests   32, 45, 49–50, 51, 56
fossil fuel combustion   41
frontal rainfall   28, 29, 60, 61

G

geographical information systems
  (GIS)   78, 80, 81
geographical skills   8
Geographic Association (GA)   78,
  80, 84
Geography Review   84
gigatonnes   **39**
GIS *see* geographical information
  systems
global biome map   47
global carbon cycle   39–44
global carbon stores   39, 40
globalisation   8
global mean surface temperature
  (GMST)   58, 59, 64,
  68, 69
global warming   14, 58
global water cycle   10–15
global water stores   11, 12–13
glucose   41
GMST *see* global mean surface
  temperature
grasslands   48–49
gravitational water   20
grazing   56
greenhouse gases   **56**, 58,
  60–61, 68
green water   11
groundwater   11
groundwater flow   17–18
groundwater store   19
gullying   **57**

H

Holocene   **53**
Hoover Dam   22, 23
Horton, R.E.   18
hosepipe bans   34
hydrographs   24–25, 27, 30
hydrological cycle   16
hydrological drought   34
hydro-topography   54
hygroscopic water   20
hypotheses   76

I

identity   8
independent investigation   7
  carrying out   77–79
  completing   80–84
  planning   71–77
inequality   8
infiltration   17, 26
infiltration capacity   17
infiltration-excess overland
  flow   18, 19, 61
inputs   10, 15–16, 39
interception   **19**, 20
interdependence   8
Intergovernmental Panel on Climate
  Change (IPCC)   58, 68, 69
interviews   79

J

jet stream   **34**

L

lag time   24
land use   26
line sampling   78
lithification   44
London Basin   37–38

M

mapping   81
marine carbon feedback   67–68
mark schemes   7, 82
mass balance   10, 39
megacities   69

meteorological drought   33
methane   39, 58, 67
Milankovitch cycles   13
mitigation   8, **51**
monoculture   **51**
monsoons   13, 30

N

negative feedback   64, 66, 67
net primary productivity   (NPP)
  40, 41, 46

O

ocean acidification   61, 68
open systems   **15**
Ordnance Survey   84
organic carbon pump   42, 43
orographic rainfall   28, 29
outputs   10, 21, 39
overland flow (surface runoff)
  17–19, 22, 25, 30–32, 61

P

pampas   48
pasture management   52
patterns   79
peak discharge   24
peak rainfall   24
peat   **39**
peatlands   **52**, 53–57
percolation   17
permafrost   **67**, 68
permeability   **17**, 18
photographs   81
photosynthesis   **39**, 41, 42, 43
phytoplankton   **43**
pie charts   81
place studies   8, 65
plant growth   47, 48
pluvial flooding   20
point sampling   78
pollution   55
positive feedback   64, 65, 66, 68
power   9
prairies   48
preceding discharge   24

Index

precipitation
    air uplift and condensation 28–30
    drainage basins 16, 17, 19
    excess runoff 30–31
    formation theories 29–30
    greenhouse gases and water cycle 60, 61
    river discharge 23, 25
    sample questions 94–97
    water deficit 33
primary data **71**, 77
processes 39

Q

qualitative data 8, 34, 77, 81
quantitative data 8, 77

R

radiation 59–60
rainfall
    air uplift 28
    drainage basins 18
    storm hydrographs 24, 25
    system and mass balance 13–17
rainforests 20, 45, 46–48, 49–50
rain shadow effect 28, 29
raised bogs 54
random sampling 78
Reducing Emissions from Deforestation and forest Degradation (REDD) 51
reliability **82**
representations 9, 79
resilience 9
respiration 41
RGS *see* Royal Geographical Society
rising limb 24
risk 9
river discharge 21–27, 31, 60–61
river regimes 22–23
Royal Geographical Society (RGS) 77, 78, 80, 84
runoff 18, 28, 30–32, 61

S

sampling strategy 78
saturated wedge 18
saturation 19, 21
saturation-excess overland flow 18, 19
scale 9
scatter graphs 81
secondary data 77
sequestration 40, 42, 51, 55, 62
slow carbon cycle 43–44
snowball sampling 78
soil moisture store 11, 19, 20–21
soil organic carbon (SOC) 51
soil texture 25, 26, 43
soil throughflow 17
solute load 43
Spearman tests 80, 81
statistical significance **80**, 81
steady-state equilibrium 12, 63
stemflow 17
steppes 48
stores **10**
    carbon cycle 39, 40, 46
    drainage basins 16, 19–21
    water cycle 10–11
store size measurements 79
storm flow 22
storm hydrographs 24–25
stratified sampling 78
surface runoff *see* overland flow
surface store 19, 20
surveys 79
sustainability 9
systematic sampling 78
systems 9, 10, 12
system threshold 64–65

T

technology 78
temperate grasslands 48–49
terrestrial feedback 67–68
thresholds 9, 64–65

throughfall 17
throughflow 17
topic frames **73**
transpiration 21, 25
tree line **68**
tropical rainforests 46–48
turf grasses 49

U

urbanisation 31

V

vegetation store 19, 20
veld landscape 48

W

wadis 18
water balance 33
water cycle
    deficit 33–38
    desertification 62
    drainage basins 15–21
    feedback 63–71
    impacts of greenhouse gases 60–61
    links with carbon cycle 58–63
    precipitation and excess runoff 27–33
    river discharge 21–27
    sample questions 94–97
    system and mass balance 10–15
water deficit
    aquifer recharge 37–38
    human causes 35–36
    meteorological causes 33–35
watershed 15
water stores 11, 12–13
water table 18, 19, 38
water transfers 13–14
weathering **43**, 44
wicked problems **69**, 70
wilting point 20, 21
WJEC examination 5, 6
WJEC/Eduqas model of enquiry 72